The New Look -

Summary

Of Summaries

ISBN 978-1-9999664-2-3

A catalogue record for this book is available from the British Library.

Published by Merops Press
Website: **www.cosmicconnections.co.uk**
www.scienceandthesoul.co.uk
www.michaelpitman.co.uk

Acknowledgements: Suzanne, Marianne, Emmanuel and Françoise.

Contents

Preface

Introducing a simple structure within which to generate
an internally self-consistent philosophy regarding information,
physics, psychology, biology and community.

Could our scientific outlook be lop-sided? Is balance needed? Is a fresh perspective possible?

This book is intended to abbreviate an explanation of what is, basically, a fresh and simple structure - **Natural Dialectic**.

Dialectic (or dialectical method) is to-fro discourse between opposing views in order to establish reasonable truth. Of course, there may exist several antagonists in a debate but, in the case of nature, it may be shown that only polarity or, at most, trinity, holds sway.

Natural, for its part, is often construed as a characteristic of any material item not devised by the mind or produced by the hand of man.

Natural Dialectic is one of different models mankind has used to try and understand the universe into which each one of us is, without asking, born. With roots deep in human thought and variously expressed at different times and places, the Dialectic's 'philosophical machine' reflects an oscillatory framework within which nature operates.

Any philosophical infrastructure, whether mathematical or verbal, is built of symbols, that is, of code or language. Language, involving a specific assignment and coherent arrangement of symbols, is the way that meaning is organised and information conveyed. In this case, could the binary structure of **Natural Dialectical Philosophy** accurately describe the *modus operandi* of cosmos? Could a spine of complementary opposites reflect the way our universe is built? Such, at least, is the dynamic formulation within which the narrative of this and parent books (see last page and website *www.cosmicconnections.co uk*) is expressed.

You can, if you wish, exploit extensive Connections/ Endnotes (indicated on the Contents page) reaching back to these volumes in order to elaborate on queries raised by The New Look's considerable abbreviation. For easy reading, devices of **bold**, underlines, *italics* and red script simply draw attention to salient points.

Illustrations have for ease been coloured according to the following themes:

light blue   universal/ cosmic

buff	informative
orange & red	axioms
purple	first causes
yellow	developmental
violet	human
red-brown	unconscious region
green	biological
silver	morality

This book is, as regards its metaphysic, scrupulously, religiously non-religious.

Finally, after careful and perhaps exciting inspection you may better judge whether Natural Dialectic's grammar well interprets nature's text and thereby accurately reflects the logic of creation. So jump aboard and ride this streamlined 'thought machine'. Intellectual seat-belt fastened, we shall travel far and fast from here…

Chapter 1: Introduction

This course is a summary - as the section called Connections that expands it demonstrates. It has **two aims**. The first involves,

in the wisdom of Albert Einstein's words,

"... *seeking the simplest possible scheme of thought that will bind together the observed facts*".

Indeed. *To introduce a simple structure within which to generate an internally self-consistent philosophy regarding information, physics, psychology, biology and community.* **In other words, to develop, by applying the razor of parsimony to all the world's libraries, an Ockham's cosmic philosophy.** At root, the idea is

to generate a philosophical routine, a 'dynamic' within which physic and metaphysic are accommodated and may be reconciled.

The second, correlated aim is to compare, as if through different angles of a prism, the logic of our material world (as explored by scientific study) with <u>the perspective that snaps into view if a single immaterial element is added</u>. What is this metaphysical addition, one very well-known and, indeed, forming the *basis* of our current age? As the industrial age was based on the energy of steam, electricity and, finally, nuclear power so ours is based on information. *Information* is **the immaterial element, that is, the metaphysical addition**.

We'll methodically compare the two perspectives[1] - call them materialism and holism - and the logic of interpretations that derive from them *using three simple models* of a balance, that is, a scale along with concentric rings and a stepped pyramid called Mount Universe.

Models are non-verbal descriptions. They are pictures to hang concepts on. To work successfully their underlying 'grammar', that is, mode of operation must be clear. We shall spend time this first lesson exploring the conceptual vehicle or 'cosmic language' which they help illustrate. If we want a fresh perspective we will need to understand this language, its mode of expression.

If correct, such language should generate an abstract, metaphysical machine, tight-knit, well-riveted by bolt and counter-bolt, the simplest working model of the universe. But what is such a vehicle called? How does this philosophical mould work? Its perspective is not mathematical; equations of a physicist may well describe the music of a song but numbers miss the whole point. Science has derived an objective, mathematical and entirely *materialistic* interpretation of events. How are facts poured into our 'dynamic' so that we add the *immaterialistic* part of our comparison and again make whole?

At this point I should straightaway scotch the notion that any considerations concerning mind, the origins of the universe or life that are not entirely materialistic are somehow **'anti-science'**. If I offer a holistic interpretation of this wall, door or the air am I 'anti-scientific'? **Am I 'anti' non-conscious matter or energy - the preserve of scientific study?** Holism deals with *both* material and immaterial elements but to say it is 'anti' one of them is absurd.

Finally, let's recall that *all* of science and philosophy is concerned to address the most natural questions of a being that finds him or herself as a body on earth, a planet in space, for a brief spell of time. Such enquiry needs answers reasoned within, I shall argue, one of two possible frameworks - materialism with its sub-creeds of humanism and so on; and holism, as old as humanity, including metaphysic *and* a science of material reality, with its sub-creeds as well. *Heterodox, we now detach from all orthodox creeds.*

> **What's it all about?**
>
> **Who exactly am I?**
>
> **Where did I come from?**
>
> **What am I doing here?**
>
> **Indeed, is there any purpose in life other than the satisfaction of animal needs?**
>
> **Then, where am I going?**
>
> **What happens, when the pot cracks, after death?**
>
> **And finally, of course, the questions of morality, law, society and government both political and personal.**

Now to business, a short but important, foundational piece concerning **Primary Assumptions**[2]. Is your primary assumption, aimed towards full truth and understanding, correctly set or not? Which answer - materialism or holism - does the evidence best brace?

What are the basic axioms and corollaries of this pair?

Materialism's axiom is that every object and event, including an origin of the universe and the nature of mind, are material alone; a few oblivious kinds of particles and forces compose all things.

Although the universe appears to work by rules and to have been established in a very particular way, this appearance of order is in fact unplanned. Its invisible framework of regulation must have occurred by chance and, since inception, individual objects and events (called actualities) occur by chance as well.

Materialism's Primary Axiom is, to repeat, that every object and event, including an origin of the universe and the nature of mind, are material alone; a few oblivious kinds of particles and forces compose all things. Moreover, cosmos issued out of nothing; therefore, beyond this realm of physics there is only void; and life is an inconsequent coincidence, electric flickers of illusion in a lifeless, dark eternity.

Although the universe appears to work by rules and to have been established in a very particular way, this appearance of order is in fact unplanned. Materialism's cosmic reason is, thereby, its own antithesis - unreason. Rules by chance, events by reflex. Oblivious matter is an aimless actor whence, by accident, you rose from the dead. Life has, hasn't it, to be the offspring of non-consciousness?

Such axiom must apply to life. In this respect the *Primary Corollary of Materialism* states, by the neo-Darwinian theory of evolution, that life forms are the product of the chemical abiogenesis of a first cell; and following that, by common descent, of a random generator (mutation) acted on by a filter called natural selection. Such evolution is an absolutely mindless, purposeless process. It is, from a materialistic perspective, a fact so that this *PCM* is a fundamental *mantra* of materialism.

Chance is the creator of diversity. Its scientific *aide-de-camp* is probability. **No matter what the odds aginst, the universe and life *must have* appeared by chance.** Order came about by, basically, chance. No telling how exactly, just vague imprecation. Nothing is, perchance, impossible; the sole impossibility is that such a story is impossible. This implacably materialistic, possibly nonsensical, narrative is rehashed in every modern textbook, journal and broadcast.

A caveat. To materialistically presume that what is not material is not natural and, therefore, cannot exist is a first order, pseudoscientific error. On the other hand, to 'pretend' metaphysic *does* exist is prejudiciously judged, by that same materialism, 'pseudoscience'. If, however, the basic nature of information is immaterial/ metaphysical then is every IT program, engineering blueprint, artistic design and, indeed, your own thinking simply 'pseudoscience'??

> **Holism's axiom is that realistic comprehension of the world includes *two* primary components - immaterial and material or, as obvious to everyone, mind and matter.**
>
> **A scientific world-view that does not profoundly and completely come to terms with the nature of conscious mind can have no serious pretension of wholeness.**

Holism's Primary Axiom is, on the other hand, that realistic comprehension of the world includes *two* primary components - immaterial and material or, as commonly perceived, mind and matter.

Holism, therefore, simply adds immaterial, as a second fundamental ingredient, to material. Or, conversely, it adds material to immaterial. Since immaterial is not material it adds nothing physical at all. **But hence follows, it is argued, this philosophy's powerful and impregnable validity.**

> **The *Primary Corollary of Natural Dialectic* states that the origin of irreducible, biological complexity is not an accumulation of 'lucky' accidents constrained by natural law and death.**
>
> **Forms of life are conceptual; they are, like any creation of mind, the product of purpose. Such assertion is, in the face of materialism, absolute anathema. *Yet, if materialism's first axiom is incomplete then every step that follows will lead further from original truth. An axiom that discounts the force of information may well be largely incomplete*.**

Holistic logic must also apply to bio-logical life. In this respect its Primary Corollary states that the origin of irreducible, codified (or highly informed) biological complexity is not an accumulation of 'lucky' accidents constrained by natural law and death. Forms of life are conceptual; they are, like any creation of mind, the product of purpose.

Such assertion is, in the face of materialism, absolute anathema. Yet, if materialism's first axiom is incomplete then every step that follows will lead further from original truth. *An axiom that discounts the force of information may well be largely incomplete.*

Furthermore, let us at the outset be completely clear - the <u>basic assertions of both materialism and holism are philosophical; *neither party is a scientific one*</u>. Holism includes metaphysic; materialism, *ad hoc*, excludes it. Material science can never prove holism's metaphysic, based on mind and information, is untrue. *In this case, if the holistic axiom that mind and matter are two different kinds of element is true then holistic logic in its entirety is unassailable.* **The previous couple of boxes express the holistic paradigm and preface its way forward.**

Three main points arise.

1. Holistic axiom exacts a toll. We need answers about:[3]

(i)   **the nature of consciousness, sub-consciousness and non-consciousness.**

(ii)   **whether individual mind can exist independent of a body and, if so, the nature of its entry, attachment, exit and disembodied condition.**

(iii)   **the interactive relationship of individual mind with body; the nature of any *PSI* (psychosomatic border or, perhaps, quantum linkage) between mind and matter.**

(iv)   **the mechanism by which universal mind, if such exists, might inform non-conscious forces, particles and gross phenomena; the origin of physical constants and patterns of behaviour, that is, the laws of nature.**

(v)   **the nature of physical and biological prototypes, homologies or, if any, archetypes.**

(vi)   **the question whether biology is informed by chance and aimless natural law or by design in accordance with such law; a wholesale reappraisal of the neo-Darwinian theory of evolution.**

2. We need to nail down the language (one not exclusively materialistic) in which holistic answers and arguments can be included - **Natural Dialectic.**

Check the preface. Any philosophical infrastructure, whether mathematical or verbal, is built of symbols, that is, of code or language. Language, involving a specific assignment and coherent arrangement of symbols, is the way that meaning is organised and information conveyed. Therefore, in order to create, understand, read or communicate messages, it is helpful to have clearly grasped whatever particular grammar is being used. So if you personally struggle with the 'imprecision' of a philosophical as opposed to mathematical structure, please keep going until until its kind of equation (or balancing) 'clicks'.

In our case the dialectical framework represents, essentially, cosmic infrastructure and is thus, simultaneously, another formulation of mankind's Perennial Philosophy. Simply, it asserts that to-fro, binary logic is the way that all things work; and is, furthermore, the way our polar cosmos is constructed. Does this assertion accurately reflect the *modus operandi* of cosmos? Does its streamlined 'thought machine' well construe the complementary grammar of polarity? We shall explore this after checking the third point.

3. **Universal models of creation.**[4]

a) You can think of cosmos as a <u>pair of scales</u>. Balance, up and down.

> **Such scale integrates a pivotal, balancing factor (*Essence*) with two antagonistic vectors[5] of** *existence.* In this, the perpetual changes of creation are seen as myriad adjustments against the disturbance of balance. Existence is a field of ceaseless change. *Creation's wheel is, therefore, one of relative uncertainty and instability.* **Such eccentric instability rolls forever perfectly imperfect; perfect imperfection shows as the perpetual motion of continual changes at local times and spaces.** Thus 'Imperfect Essence' (called existence) sums to swinging of a single but self-regulating balance. Its equilibrations generate the myriad individual actions and reactions each of which always shows proportions of two vectors pivoted around the central hub of a third non-vector.

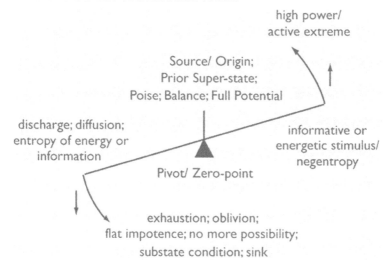

Now inspect this triplex 'stack':

11

↓ down/ descent	Pivot	up/ ascent ↑
sink	Source	flow
gravity	Equilibrium	levity

Note, for future reference, trinity, the pivot and the vectors (up and down); capital letters for the central column; and flow (or relative between Source and (relatively inactive) sink.

b) You can think of cosmos in *dynamic* terms of <u>**concentric rings**</u>.

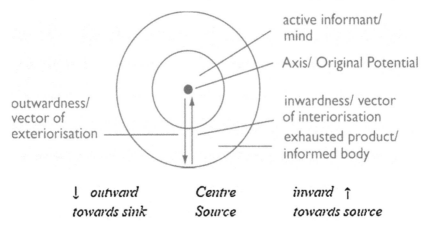

↓ *outward*	*Centre*	*inward* ↑
towards sink	*Source*	*towards source*

This figure of concentric rings includes antagonistic vectors to and from its Central Source. Vectors run inward (towards) this Source or outward (from it) towards peripheral sink. With this model we are simultaneously building a picture of spectrum, that is, scale of energy or information. **Natural Dialectic calls it the conscio-material gradient, or spectrum, of creation.**

c) Third is the idea of <u>**Mount Universe, the Cosmic Pyramid.**</u>

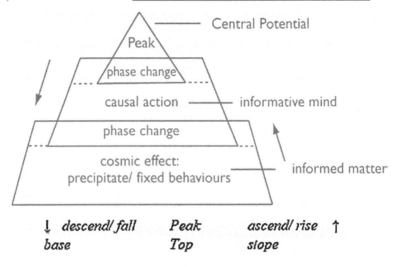

↓ *descend/ fall*	*Peak*	*ascend/ rise* ↑
base	*Top*	*slope*

12

A cone or, squarer, pyramid describes '*static*' hierarchy. A useful representation is the stepped pyramid, also called a **ziggurat**. In this case each step of a **ziggurat** stands clearly for a phase, level or stage; and the apex, its capstone, a point that points beyond the finite grades below, implies peak infinity. **This capstone is the highest point and source of what we call Mount Universe**.

Models act as visual metaphors. Of these three you can use which one best illustrates a point you're making. Their Source, Centre, Axis, Peak and Pivot are essentially equivalent.

And up and down (regarding motion or transformation) mark the basis of existence called, in other word, **duality**. Change and relativity are the nature of all things. But it is obvious, from the triplex stacks, that we are also dealing in **trinities**. The nature of this Central, Balancing Source is also, as we'll see, of great interest.

(-) *negative*   *Neutral*   *positive*   (+)

Positive, negative and Neutral. The words in stacks are not arranged haphazardly. Their order represents a logic. Triplex by triplex, stack by stack we can and shall build a simple but a self-consistent picture of the world.

Moreover, three components of each model, two vectors and a point of balance, find various expression in each object and event in creation - even, we shall find, in description of creation as a whole. More subtly, they reflect operating principles whose permutations are expressed, in varying degree, as *tendencies* both psychological and physical. **Since the triplex is all-pervasive to the extent of describing the three major tiers of Mount Universe itself, we call them <u>cosmic fundamentals</u>.**[6]

If the triplex is naturally fundamental it is also fundamental to the systematic operation of the Dialectic we are formulating and can now begin to use; but while the orient has long worked and woven with these immaterial radicals, these fundamental threads, then western minds have not.

↓   *yin*   *Tao*   *yang*   ↑

You might be familiar with this far-eastern, Taoist triplet and its associations. Therefore, to head its stacks, Natural Dialectic uses an equivalent, equally ancient abstraction.

↓   *tam*   *Sat*   *raj*   ↑

Sat **(equilibrium),** *raj* **(stimulating ↑) and** *tam* **(inertialising or materialising tendency ↓).** We're almost there. The **first of three final**

steps is to parse these triplex stacks as dualities. **This is because Dialectical stacks are a columnar expression of polarity,[7] that is, duality.** Let's start with a simple 'duplex' presentation of the vectored parts. It is called a stack.

↓ *fall*		*rise* ↑	
(-) *negative*		*positive*	(+)
exhaustion		*stimulus*	

A *stack* is a set (or pile) of members. Each *member* of a stack (e.g. negative *and* positive) consists of a pair of polar 'anchor-points'; and each *element* of the pair (negative *or* positive) is arranged according its fundamental characteristic viz. its tendency to (*raj* ↑) rise or (↓ *tam*) fall. Health/ illness, life/ death, happy/ sad - the list is endless.

It is immediately obvious that members of a stack are not synonymous. *But they equate in fundamental character; they align in terms of cosmic fundamentals.*

↓　*dark*　←　　　→　*light*　↑

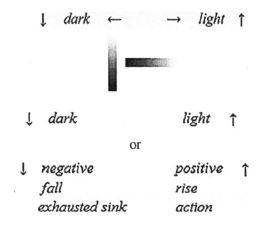

↓　*dark*	*light*　↑

or

↓　*negative*	*positive*　↑
fall	*rise*
exhausted sink	*action*

Also, each member implies a scale of greys or **spectral range** that runs between its elements, that is, its '**paired opposition**' or '**complementary covalency**'. For example, you can have 'more or less negative'; you may suffer a 'greater or lesser' fall. *As a scale of greys runs between extreme black and white, so it is implied that a scale permits oscillation of values between any pair of elements.*

On any scale motion runs down (↓) and (↑) *vice versa*. **Thus a single, non-repetitive first-line representation elegantly indicates the direction of elemental vectors for the whole stack.**

What about the central (*Sat*) character? Sometimes it is more convenient to write a non-vectored di-logical stack and then polarise its neutrality into a second (*raj/ tam*) vectored and also di-logical membership. **What is meant by this? How is it done?**

$\downarrow$ (-) *negative   Neutral   positive*   (+) $\uparrow$

This triplex is resolved into two separate stacks. The first, top stack of the pair disposes Neutrality on the right against a neutralised description of the two polarities on the left.

$\downarrow$ *polarity* $\uparrow$ *Neutrality*

The left-hand element is then split into its polar vectors, ($\uparrow$) positive and ($\downarrow$) negative, in a second, vectored kind of stack:

$\downarrow$ *tam*	*raj* $\uparrow$
negative	*positive*
divisive	*unifying*
fixity	*flux*
lock-up	*freedom*

As was mentioned, for Natural Dialectic the ceaselessly moving, changing forms of creation are called existence. Their changeless Source is called Essence.[8]

$\downarrow$ *existence* $\uparrow$ *Essence*

We can include this member, with or without vectoring arrows, in a slightly larger build:

existence	*Essence*
relativity	*Absolution*
polarity	*Neutrality*
duality	*Unity*
motion	*Inaction*

Such a binary stack is called Primary, Essential or Central Dialectic. *It is indicated by writing the right-hand, Essential column with a capital letter.*

A Primary Stack sets (*Sat*) Unity against ($\downarrow$ *tam/ raj* $\uparrow$) duality; or, if you like, it sets qualities of motion against Inaction or relativity against Absolution.

$\downarrow$ *tam*	*raj* $\uparrow$
negative	*positive*
exclude	*include*
isolate/ specify	*generalise/ link*
materialisation	*dematerialisation*

Duality, however, implies polarity. **Such polar component is expressed in the lower, vectored so-called** secondary, existential or peripheral/ polar dialectic. Thus **secondary, existential stacks**

15

(*written exclusively in lower case*) represent the various kinds of polarity from which the changeful web of existence is composed.

These two kinds of stack can be combined to obtain a full picture, in any circumstance, of the three fundamentals. The combination is called a linked or full stack.

Primary, Essential or Central Dialectic:

tam/ raj	*Sat*
existence	*Essence*
polarity	*Neutrality*
expression	*Potential*
limitation	*Infinity*
duality	*Unity*
relativity	*Absolution*
motion	*Balance*
something	*Nothing*
(3)/ (2)	*(1)/*

secondary, existential or peripheral dialectic:

↓ *tam*	*raj* ↑
fall/ down	*rise/ up*
negative	*positive*
division/ multiplication	*unification*
isolation	*connection*
drag	*stimulus*
(3)/	*(2)/*

In the above-mentioned example and the two that follow all elements of the right-hand Primary Stack reflect the character of Essence. Essence is Supreme Being without predicate. It is Uncaused First Cause, the supra-existential root of creation. From Unimaginable Essence issue all objects and events, all changing forms. (see endnote for links).[9]

We now know how to construct a linked or binary pair of stacks. As will be shown, this can be done for any different context you may choose to clarify. In some cases, the logic may force us to change muddled preconceptions.

Reading down any column connections, some obvious, others less so, appear within the pattern. For example, Infinity,[10] Unity and Nothing are not synonyms but, paradoxically, appear to represent aspects of Essence. Of course, we build on identifying the character of Essence and of existential relativity throughout the book. In some cases, the logic may force us to change muddled preconceptions.

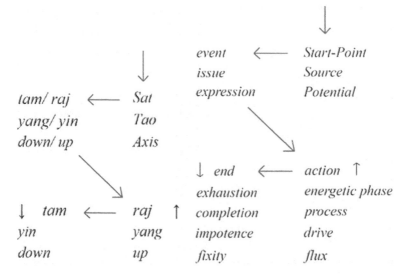

A good, *top-down* way to read a stack is, as in these two diagrams, to start with the arrow above right of the Primary Dialectic's right-hand column; then track to its left-hand, polar column; from there move to the active (*raj*) side of the secondary stack; and, finally, to the left and down.

Start, action, end. From source (potential) through issue to exhausted sink. From start through process to completion. Every object or event involves prior precondition (called potential), an active phase and a passive, finished or fixed phase. This, as we saw earlier, includes not only the course of all events, great or small, but the process of creation itself.

At this point, therefore, in **the second of three final steps** we need to resolve a paradox viz. the ubiquitous reflection of Essential characteristics in motionful existence. We need to discriminate between characters on the right-hand Essential stack and their limited temporal appearances. For example, balance, unity and potential occur at myriad times and places in existence. And Source transcends mobile creation; it is Cosmic First Cause but is, paradoxically, reflected endlessly in lesser psychological and physical causes. We distinguish, in effect, between Absolute Nature and its reflection in finite phenomena.[11]

In short, the characters of the Essential column are commonly *reflected* as part of creation *but only in local, temporary forms*. Every existential object or event has boundaries; it is - psychological (in the subjective case of mind) or physical (in the objective case of bodies) - conditioned. While Infinite Reality is unconstrained we call constrained, differing and relative realities '*lesser*', '*seeming*' or '*apparent*'. They are *phenomenal* and finite in more or less degree.

Time and space are interesting *lesser infinities*. Other phenomena (including balances, potentials and so on) are relatively restricted, that is, localised in time and space.

Now for the **third step - inversion.**

existence	*Essence*
issue	*Source*
motion	*Inaction*
↓ *inaction*	*action* ↑
anti-source/ sink	*current*

In the stack above note that 'inaction' occurs in both top-right and bottom-left columns. The word is the same but its nature completely different - as different as source from sink. Similarly with the word 'peace' - potential (no action yet) and RIP (action over). Such spiral' down from pre-active to opposite, post-active condition is called **reflective asymmetry** or **inversion.**[12]

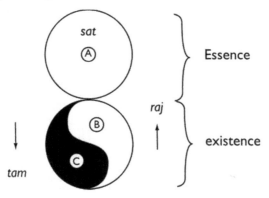

A binary system intrinsically includes inversion between opposites or, if you like, from top to bottom of a stack; or, in the diagram, from A to C. It is the way creation works. As regards cosmos (and, therefore, your microcosmic self) such inversions run from the expression of Informative Potential to its complete absence (or locked, impotent fixity) in non-conscious, automatic matter; and, as regards energy, a fall from subtle quantum elements to the gross, locked expression of solid matter. In other words, two kinds of inaction - that of pre-active, poised potential and post-active, exhausted finish - are asymmetrical reflections of each other. They are, as are also the unchanging, general stability of precondition (in physics called law) and the changed, specific stability of a fixed outcome, entirely different. These, informative and energetic, are the couple of *Primary Reflective Inversions* that, interlocked, compose the cosmic conscio-material gradient of creation or, if you like, the informative/ energetic scale of cosmos. Such gradient is no figment; for example (as Chapter 5 illustrates) your own physical structure reflects it.

We are now, at last, in a position to ask, "How is all this related to science?" [13] **How can stacks and cosmic fundamentals re-angle not the facts but the prism of our perspective?**

In fact, we can see and will elaborate the triplex logic within which we might profitably and comprehensively treat the scientific disciplines of (mind) psychology, matter (physics) and their connection, mind-with-matter, in biology.

> *A simple link from <u>psychology</u> describes the three basic conditions of information as (sat) potential informant (concentrate of consciousness) prior to (raj) active mind (both informant and informed) and (tam) passive, dormant or sub-conscious mind.*

> *A simple link from <u>physics</u> describes the three basic conditions of energy as (sat) potential prior to (raj) action and (tam) exhaustion.*

> *Thirdly, a simple link from <u>biology</u> describes the three basic conditions of life on earth as (sat) informative archetype, (raj) metabolism including a code carrier called DNA and (tam) finished product called developing or adult body.*

In other words, there exists a holistic, triplex (threefold) treatment of science as well as the one-tier materialistic one; and that, as regards holistic logic's fundamental connectivity, there are thousands of possible stacks. I have used just an exemplary handful in this chapter. *What they amount to is a system for generating fresh perspective, links you have not made before, new insights into cosmic connections.*

To summarise: **Natural Dialectic asserts that to-fro, binary logic is the way that all things work; and is, furthermore, the way our polar cosmos is constructed. Such radical infrastructure involves a trinity of qualities called cosmic fundamentals.**

The philosophical structure, which includes an immaterial element, information, represents **holism**. At this point we note that current science by default frames its explorations in a materialistic, evolutionary framework. *Thus to reappraise the data in a holistic manner amounts to fundamental change in a basic concept of scientific discipline.* **This is what physicist and philosopher Thomas Kuhn called a <u>paradigm shift</u>.**[14]

Henceforward we'll see how the fundamentals fit with science and with the construction of Mount Universe. In the next chapter we investigate the question of immaterial information.

Chapter 2: Information

Let's straightaway, having outlined the structure of our philosophical vehicle, throw down the gauntlet. Let's enter the halls of scientific discipline announcing three great issues we'll address.

3 Great Issues of Science and Philosophy

Psychology: The Nature and Origin of Consciousness

Physics: Cosmogony, The Origin of Physicality

Biology: The Origin of Codified Life-Forms

In this exploration we draw a distinction between two basic modes of enquiry both elaborated in the section on psychology. The first is **subjective focus** - contemplation which allows planning and perception of principles; the second is **outward focus** - perception by means of sensation, muscular manipulation and each's technological extensions. These 2 modes of information gathering and processing will be discussed later in the section on psychology. For now, simply note that these ***anti-parallel vectors*** of mind correspond to what we call *bottom-up* and *top-down* directions.[15]

Bottom-up is, broadly, that of the person working from detail to understanding patterns and principles by experience. This kind of logic, **from detail to principle, is called inductive**. It is the way of naturalistic curiosity, that is, of experimental science.

Top-down is, broadly, that of the expert working **from principle and its application for making sense of details.** This kind of logic, **from principle to detail, is called deductive**. It is the way, from mind to matter, that holism works.

There is a third, important sort of logic that we'll come across. **It applies to unique historical events or futuristic speculation that we have neither seen nor can for sure repeat. It is called abductive reason or conclusion by best inference**.

Now, in dialectical respect of principle, I'd like to draw attention to duality within unity and the basic elements of the existential dipole drawn below.[16] We will shortly relate its triplex nature to Mount Universe. First note that the two basic elements composing all the various forms of creation, that is, of existence are information and non-conscious energy. In each exists a scale of informative and energetic phases we identify as, respectively, mind and matter.

Materialism has decided matter is the basis of creation (though it does not specify the nature of the nothingness that matter must have started

20

from); thence mind is an excrescence from a brain and soul a pure (but physical) imagination. On the other hand, holism's revolution turns materialism on its head. The diagram below shows The Monopole, the nature of Sole Cosmic Independence - Infinite Essence; and its dependent existential dipole within which occur all motions of creation.

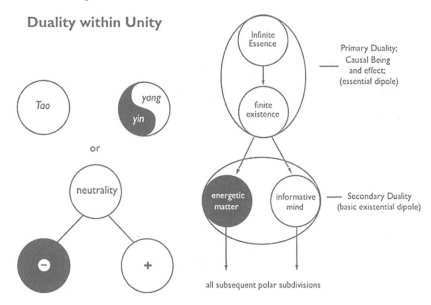

We are now in a position to relate cosmos in terms of **the triplex yin-yang-*Tao* model and the cosmic pyramid: (*Sat*) Essence with (*raj*) mind and (*tam*) matter**.

Cosmic Fundamentals and their Ziggurat

Three Tiers of Mount Universe with Sub-divisions

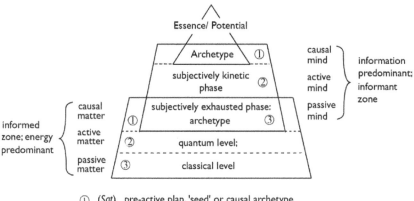

① (*Sat*) pre-active plan, 'seed' or causal archetype
② (*raj*) internal informant, pattern-maker; primary effect
③ (*tam*) external structure, fixity of pattern; secondary effect

21

tam/ raj	Sat
existence	Essence
expression	Potential
issue	Source
↓ lower	upper ↑
physic	metaphysic
non-conscious	conscious
reflex/ passive	active/ creative
material	immaterial
body	mind

This ziggurat, stack and the table below help to elaborate the previous diagram.[17] Together they make clear the triplex nature of cosmos in terms of the three cosmic fundamentals and the basic existential dipole. The whole is in three - (*Sat*) Source, (*raj*) mind and (*tam*) matter. The latter pair, are also divisible into sets of three. We'll deal with information in terms of its triplex sub-divisions later in this chapter and, similarly, with related scientific disciplines in chapters 3-6.

Upper Pole - Information
   (Sat) Potential Information
   (Raj) Active Information
   (Tam) Passive Information

Lower Pole - Energy
   (Sat) Potential Energy
   (Raj) Active Energy
   (Tam) Passive Energy

Before moving on, please relate the phase of *archetype* (see the above ziggurat and Glossary) with active, causal super-matter *and* passive information, a memory file. Archetypes[18] constitute the *potential* phase of both psychological and physical triplex. As such archetypal source is an important feature of holistic cosmos. For example, think of passive archetype as a 'seed' that precedes yet governs the origin of physicality (say, a big bang); you might allow that the precedence of its operations would follow different rules from those in space and time. Could it exist, like a kind of informative '*DNA*' in the holographic body of the universe, apparently nowhere - being metaphysical - but actually everywhere and every time at once? We'll make a further check in Chapter 4, the section on physics.

Every scale or hierarchy has extremes.[19] Now let's take a brief look at the ziggurat's extremities, that is, the thesis of its extreme initiation and, at the other end, consequent extreme antithesis. At base rests non-conscious matter, a universe whose ultimate sinks are perhaps the massive, local 'infinities' of black hole singularities. The opposite of

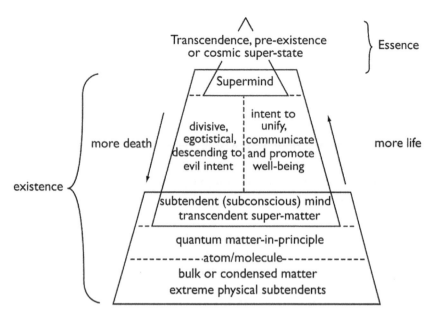

non-conscious is conscious. In this case we identify, at immaterial super-state, the ultimate, uncaused source as the Singularity of Consciousness. This formless initiator is, logically, the extreme concentrate or purity of consciousness.

In other words, uncaused and eternal Transcendent Super-state is regarded as the Centre, Source or Pivot round which cosmos swings; and the sink/ periphery is non-conscious matter.

Why, though, put consciousness above non-consciousness? Materialism asks why *eternal matter* of some species shouldn't be the uncaused source of all there is. We'll deal later with why science thinks all physical phenomena were actually caused, material cosmos had a start and steady-state theories of eternal matter have been binned.

Conversely, why should Super-state hierarchically precede mind as well as well as matter? **Why should it more likely be, as we shall see, the uncaused Cause of Everything?**

It is true, *top-down*, that products subtend their producers. A producer stimulates and stirs things up. Cause precedes consequence. *From this perspective active informant precedes its passive, informed consequence, metaphysical mind plans physical arrangement and super-natural precedes natural order.* **Mind first, body after; body's an appendage of its mind.** It happens all the time with us: idea (or desire) precedes outcome. *Does it, that physical depends on metaphysical, invert your 'normal', that is, 'sensible' sense of things?*

So, for example, we say that sleep subtends waking, sub-conscious subtends conscious. And, conversely, waking transcends sleep. It is like the various bands of a spectrum - UV, visible (us) to IR and black.

We're emphasising a notion of *hierarchy* central to a tiered universe. Spectrum introduces an idea of hierarchy, scale or, in ancient parlance, Jacob's ladder.

Hierarchies always have source. A boss. A cause. ***So now the hierarchical interest turns to causes.*** Of course, we can think of causes in terms of knock-on effects. They impact, in physics, chemistry and psychological reflexes, in '*horizontal*' chains. **No doubt, energetic causes push effects; they bump you from behind; their arrow, physic's arrow, runs from <u>past to present</u>.** And things suffer (though you'd hardly think it as regards a proton or electron) from increasing weariness called entropy. They run out of steam. We call this <u>horizontal causation</u>.

What, though, about a cause that is conceptually implanted? What about not energetic but *informative causation*? **This is goal-oriented; and goals are <u>in the future pulling you their way</u>. They pull you from the future; they lift forward. They are metaphysical *attractors*, guides that govern your behaviour as they lead you through the world.** Not material but immaterial, such leadership is not by force of gravity, electric charge or magnetism; information is metaphysical not physical; it's psychological and, although at this point you may want to know exactly how, be patient - as the course elaborates we shall come to see. *Information's entropy is negative.* **Mind is negentropic and thus metaphysic's arrow flies, from future back to present, anti-parallel to physic's.** Thus you are guided, present to the future, by plans or programs that realise their goals. For example, take the rules of football - unseen and unexpressed until a game is played.

Waggle your little finger. What *caused* what? The line of operational command runs from conscious, informant mind through sub-conscious templates (memories, instincts and so on, dealt with later in Chapter 3) over a psychosomatic border to to innervated, muscular body. Between the zones of purpose and non-purpose, incoming or outgoing information would be first translated to, or last translated from, the chemo-physical side by the product of 'excited' electrons, that is, electromagnetism. Where *matter subtends mind*, this electronic phase is in turn subtended by the biochemical; and both levels occur within bulk, biological structures, that is, cells, nervous system, muscles and a coherent whole body. **This places your finger-waggle at the base of an informative/ energetic hierarchy.**

This process we call <u>vertical causation</u>.[20] **It is the order of an act of creation.** Its 'phased intent' drops from creative mind to created material form. As a later talk will elaborate, your own sensible, physical form (perhaps including the origin of its shape) resides at the bottom of such a hierarchy.

The Order of an Act of Creation

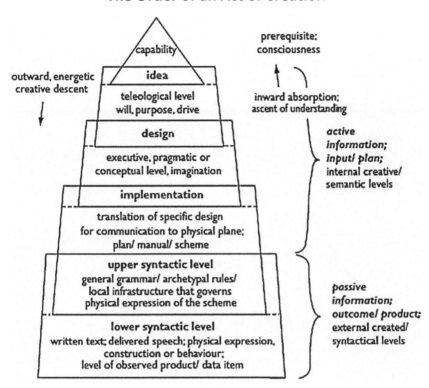

In the previous chapter we learnt how to read a full (Primary and secondary) stack from top to bottom; and in this chapter it was noted that start (or source) through mind to finish (or material end-product) can be portrayed as three tiers of Mount Universe.

We are now in a position to elaborate this 'reading routine' to simply define, using two full stacks, <u>the order of an act of creation</u> as regards cosmos itself (cosmogony with its product, studied by cosmologists) and any single, individual event within it. At the same time these Essential and existential routines demonstrate both the hierarchical and cyclical nature of creation.

It is useful to rephrase such a new and powerful routine. The logic of Natural Dialectic indicates a drop between immaterial and material extremes, that is, from pure concentrate of consciousness to pure non-consciousness (the state of matter). It thereby suggests a 'conscio-material gradient' of cosmos, that is, a spectrum or hierarchical structure based on level of consciousness. In this hierarchy Immaterial First Cause is the Zero-Point; it is the Origin, a Source called Essence.

Such Universal Initiation, giving rise to the cycles of existence, is illustrated in the first diagram.

Essential Cosmos

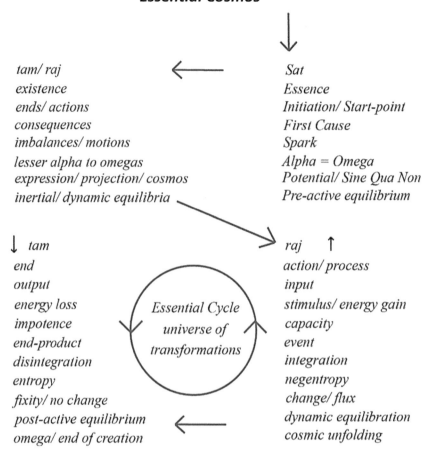

tam/ raj ← *Sat*
existence *Essence*
ends/ actions *Initiation/ Start-point*
consequences *First Cause*
imbalances/ motions *Spark*
lesser alpha to omegas *Alpha = Omega*
expression/ projection/ cosmos *Potential/ Sine Qua Non*
inertial/ dynamic equilibria *Pre-active equilibrium*

↓ *tam* *raj* ↑
end *action/ process*
output *input*
energy loss *stimulus/ energy gain*
impotence *capacity*
end-product *event*
disintegration *integration*
entropy *negentropy*
fixity/ no change *change/ flux*
post-active equilibrium *dynamic equilibration*
omega/ end of creation *cosmic unfolding*

Essential Cycle universe of transformations

There exists, equally, existential initiation. This simply means a start-to-finish of every single event in a continually mobile cosmos. This case is illustrated in the second diagram called 'existential event'.

Every event cycle, succinctly expressed as causal stimulation yielding action that dies away to an end-point, is recycled with a fresh stimulus. In this, the cause (or potential) for any event may be vertical (psychological) or horizontal (by physical reflex or knock-on effect). In other words, the second illustration, existential cycles, includes the vertical causation or psychological events of purpose/ desire; and the horizontal causation of physical events, that is, mindless, reflex interactions ('passive information' in the previous yellow illustration).

Events can happen simultaneously within each other e.g a cloud in the sky or atomic oscillation within molecule within larger body. It is not nature's but an observer's problem which event is singled out for focus.

Regular cycling/ recycling in space-time is called vibration. The oscillation round a norm of such dynamic equilibrium is called a wave.

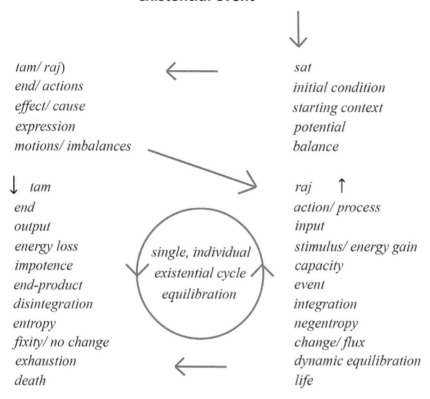

existential event

tam/ raj)	*sat*
end/ actions	*initial condition*
effect/ cause	*starting context*
expression	*potential*
motions/ imbalances	*balance*

↓ *tam*	*raj* ↑
end	*action/ process*
output	*input*
energy loss	*stimulus/ energy gain*
impotence	*capacity*
end-product	*event*
disintegration	*integration*
entropy	*negentropy*
fixity/ no change	*change/ flux*
exhaustion	*dynamic equilibration*
death	*life*

single, individual existential cycle equilibration

And a dramatic example of *irregular* cycling is an explosion. Preconditions must be correct; there follow stimulus (detonation) and the randomising decay of fall-out until motionless silence, the end.

Put another way, existential cycles, whether psychological or physical, amount to causes and effects within the regulated context of animate and inanimate conditions. They may be mindless and automatic or the outcome of purposeful intent. For example, a machine with cog-wheels systematically coordinated to produce a desired operation (such as a clockwork watch) is a specifically designed outcome. A clock is many orders less complex than any bio-machine, that is, biological body. Whichever way, mindless or mindful, many other kinds of cycle such as planetary, temporal, ecological and bio-cybernetic are familiar to everyone. So are life-cycles and, from idea to fulfilment, the cycle of desire. Existence runs around in circles.

As you have seen the holistic view holds vertical cause *higher* than its effect. The finger-waggle demonstrated how you personally operate. However, we can reach beyond the normal cycles of existence both psychological and physical. We now focus on Source, Centre or Highest, All-Informant First Cause.

Aristotle believed in a First Mover itself unmoved by any cause. St. Augustine observed that no 'efficient cause' can cause and thus precede itself. Thus causal order can't be infinite; there needs to be an uncaused primal cause. Existence is composed of caused, finite events. Whatever begins to exist, asserted the *sufi* Algazel, has a cause; *and 'something which begins has a sufficient cause' is also the modern principle of causality*. **This principle is constantly verified and never falsified. The physical universe began to exist and therefore has a cause**. What is caused is not eternal. It is finite. Its effect becomes a further cause. Thus all existence is a changeful network made of causes and effects; creation is an action and reaction zone.

In summary, it's as simple as it's crystal clear. What starts to exist is always caused. We presume the physical universe started and eternal matter, that is, endless change, is not a feasibility. Therefore, material cosmos, known as nature to the natural sciences, started to exist and has a Natural Cause. *This cause, preceding physicality, is physical non-being.* It is nothing physical but causes cosmological effects. What came (or comes) before the latter's matter, space and time must itself be time-less, space-less, immaterial. Preceding its secondary causes and effects the primary, first cause must be physically uncaused; it is uncreated in a naturalistic sense, thus super-natural. Metaphysical.

What, therefore, is the Nature of Informant Metaphysic?

Primary (Metaphysical) and Secondary (Physical) First Causes[21]

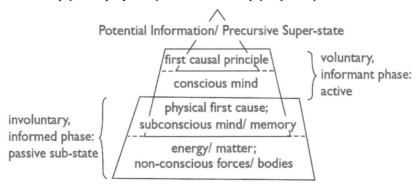

The first thing to note is that, in cosmic terms, *physical first cause is **not** the same* as *Psychological First Cause* (or Pre-motive, Essential Super-state). In short, the following ziggurat and causal cones illustrate two very different kinds of first cause, the **higher conscious** and the **lower unconscious**. Their natures will be elaborated in the following chapters.

Now, after this preamble, it is time to focus on the nature of Natural Dialectic's immaterial element, information.[22] We are information junkies. We want the news, crave more information and want to know everything! Perhaps not only externally in the universe (including our bodies) but on the immaterial, contemplative mind side

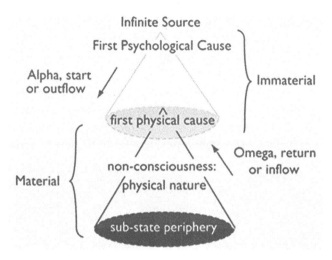

too. So what is info? It is defined as facts or fictions learned, knowledge, understanding or intelligence (e.g. sigint); also as computer data or what is conveyed by a particular arrangement of things or symbols. The arrangement of symbols, representations or stand-fors is called code. Language. **But is there any kind of information without mind?**

tam/ raj	*Sat*
below	*Transcendence*
range of forms	*Super-State*
output/ event	*Pre-motive Potential*
matter/ mind	*Informant*
↓ *tam*	*raj* ↑
passive	*active*
matter	*mind*
passive information	*active information*

Top-down, to receive and signal information is the immaterial gift of mind and never matter. **It is irreducible to scientific scrutiny. Yet, as we'll see, its 'semiotic' metaphysic dominates debate about creation and our lives.** Pick up a postcard, menu, letter - anything informing you. The object is reducible to chemistry and physics but the *sign* or *message* it conveys is not. Signs and signals always have a purpose; objects *per se* never do. In other words, information is fundamental to our being but does not fall, in a semantic sense, within the scientific remit. *Particularly, mindless origin of bio-code is an irrational hypothesis.*

Bottom-up, **however, everything is seen as energetic interactions. Energy's the physical informant.** Information's therefore, even in the case of brain, evolved by chance and natural law. **Oblivious physic generates its off-chance but never has a goal; locked in 'horizontal',**

reflex causal cycles matter mindlessly churns transformations. *Particularly, knock-on origin of bio-code is, from this perspective, a rational hypothesis.* Is such metaphysical consideration an illusion, a confusion or the truth?

Firstly, in this dilemma of perspective, let's distinguish mundane (normal, meaningful) sense from scientific/ materialistic understanding. *In mathematics of the latter sense 'information' must not be confused with 'meaning'.* So what exactly, in this apparent shortfall, *does* the word mean?

Claude Shannon   Norbert Wiener

It was Claude Shannon who, with Warren Weaver, first devised a mathematical and thereby scientifically acceptable definition of information. Shannon treated its transmission in purely physical terms according to statistical formulation of the entropy-inclusive laws of thermodynamics (2nd Law is perhaps most basic and tested in physics, no exception ever observed). In such transactions his unit was the *bi*nary digi*t*. This on/ off, one-zero 'bit' allowed the quantitative properties of strings of symbols to be formulated. His theory inversely relates information and uncertainty. The more uncertain, the less probable a sequence of symbols or arrangement of materials is, the more information it is calculated to contain. Rephrased, an amount of information is inversely proportional to the probability of its occurrence by chance. Simple, repetitive or predictable sequences contain less and complex, irregular arrangements more information. **Thus 'Shannon information' is a simply measure of improbability.** *It makes no judgement whether such irregularity is specified; it involves no sense of meaning.*

ZNQW&NSIXT AZ2NVB
COMPREHENSIBILITY

Shannon's definition is suitable for describing statistical aspects of information such as quantitative aspects of language that depend on frequencies (e.g. as how many times the letter 'a' or the word 'and' occurs) but it treats any random sequence of symbols as information without regard to its concept, meaning or purpose. In other words, the more improbable any arrangement the richer is its Shannon-defined information content. *In short, two messages, one meaningful and the other nonsense, can be exactly equivalent according to this form of analysis.* For example, ZNQW&RSIXT AZ2HVB and COMPREHENSIBILITY are assigned the same value. In other words, Shannon has reduced information to a statistical quantity; he has shaved off any sense of meaning, cut out sense of purpose in numerical analysis. *Shannon-shaving deals the same with purposive (vertically caused) and non-purposive (horizontally caused) complexity.*

In this mode of thinking a 20-letter randomly-generated and meaningless sequence contains information richer than the simple phrase 'I love you'.

It is important to grasp how Shannon, thoroughly negating logical reality, accords randomness and purpose equal status. But randomness is really reason-in-reverse; it is information's opposite. So when it comes to language or to mechanism, including those embodying biology, such conflation is an error of first order. Shannon's analysis does not distinguish between presence of mind, authorship or creativity and their absence; it fails to recognise purposive specificity; nor does it accord function or meaning any premium. *Thus order and precise meaning - usually complex, always accurately coded - might as well be able to emerge from randomness or senseless motion under natural laws of physics.*

Shannon's definition involves counting the frequencies of physical events; and physicists, measuring non-conscious energies, sometimes refer to 'information-carrying' interactions. However information is, like energy/ matter, a fundamental quality as well as quantity. Science investigates the **passive**, lower syntactical phase of information whose causes inform horizontally. Such energetic interactions are mindless; the only information involved is, extricating principles and patterns, in an observer's mind. Ignoring this, however, you may fool yourself and think in topsy-turvy terms of 'senseless design', an anti-teleology well-known as Darwinian evolution (Chapter 5).

Informative capacity inhabits the *arrangement* of material - not just any old arrangement but one with meaning that serves purpose or accords with principle. For example, everything around you in your room, not least the simple chair you are sitting on, involves informative capacity. It involves the **active** quality of information, mind. Purpose weighs no more than understanding; both weigh just as much as meaning. In the scientific balance meaning weighs as much as abstract theory. Each is lighter than a feather. Universal mind would weigh precisely nothing yet remain a vital hidden variable. *If, though, information is absent massively, what is it?*

If immaterial then, this volume shows, information renders materialism with scientism (but not science) most illogical! Indeed, Norbert Wiener, mathematician and founder of cybernetics and information theory, said, *"Information is information, neither matter nor energy. Any materialism which disregards this will not survive one day."*

Norbert Wiener, mathematician and founder of cybernetics and information theory, said *"Information is information, neither matter nor energy. Any materialism which disregards this will not survive one day."*

Information is not a property of matter. Purely material processes, unguided except by natural law, are fundamentally precluded as *sources* of information. That is to say, there is known neither law nor process nor sequence of events through which oblivious matter can create or collect information.

There is, furthermore, known neither law nor process nor sequence of events through which oblivious matter can create or collect information. **The latter is not a property of matter.** Purely material processes, unguided except by natural law, are fundamentally precluded as *sources* of information. Information is not a thing itself but a representation of physical things and metaphysical entities. Physical nature wears no numbers, bodies are devoid of words; these are planted or taken away by mind.

To rephrase, information involves the transformations and arrangement of material phenomena but such phenomena *per se* are purposeless; they can inform a mind (by sense, mathematical descriptions etc.) but not their own aimless, mindless, non-conscious oblivion. However, although the default materialistic position is to claim laws of nature operate within a field of chance encounters, it is possible to holistically argue that physic provides a dynamic stage for purpose, that is, for life. *There may be a purposive source (or program) that informs specific, automatic behaviours.*

In this case, let's take a second angle on the way that cosmos and the human individual are informed.

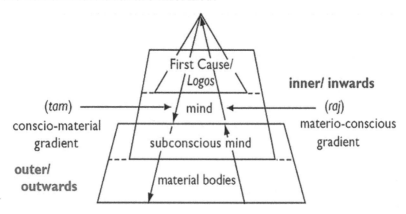

This is a key illustration. Its ziggurat reflects **the information**

gradient from active, inner to passive, outer phase; and also the diagram showing 'Two Kinds of First Cause'. As such, it illustrates the anti-parallels of a dynamic conscio-material gradient that run up-in or down-out. This drops from Clear Formlessness at peak through psychological and physical forms of various quality. The latter are trapped in reflex, oblivious cycles; the former involve, at least for humans, a measure of choice.

For '**inside information**' you can see the way to go; and how, conversely, sensation leads you 'down' and 'out' into the world of '**outside information**', that is, of physical events. You can thereby easily understand why the sage turns his contemplative focus inwards towards unification with First Cause; and a scientist outwards to the details and, in principle, unification of an essentially non-conscious universe. Sharpened focus. Concentration does the trick. One man could incorporate both sage and scientist!

X, where you live, is the tipping point between the inner and the outer worlds. X, the 'third eye' of thought, is your micro-cosmological axis. Called mind, it is the information centre where you live. Now, with regard to macro-cosm too, you have your bearings.

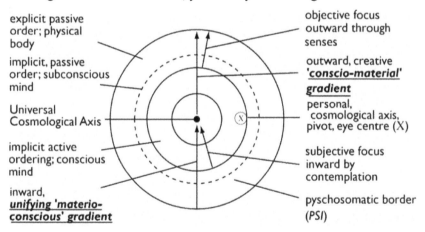

explicit passive order; physical body

implicit, passive order; subconscious mind

Universal Cosmological Axis

implicit active ordering; conscious mind

inward, **unifying 'materio-conscious' gradient**

objective focus outward through senses

outward, creative **'conscio-material' gradient**

personal, cosmological axis, pivot, eye centre (X)

subjective focus inward by contemplation

pyschosomatic border (PSI)

Where peak is central axis, the diagram below is another way to understand the previous key illustration. *And if we refer back to an illustration called The Order of an Act of Creation we can combine all three models to demonstrate, as follows, the triplex nature of information according to the fundamentals sat, raj and tam.*[23]

First in this triplex issue comes (*Sat*) **Potential Information** at the central cosmological axis or source of mind. Such conscious, inside information is what the contemplative sage is fundamentally after. Look *top right* where a sage's hand is pointing. It leads up towards Absolute, as opposed to relative, Truth (see also Chapter 6). It is experienced as what is called Illumination.

restricted consciousness	Illumination
mind	Essential State
relativity	First Cause/ Logos
relative impotence	Potency/ Control
↓ unconscious sub-state	mental spectrum ↑
darkness	range of shades
total lack of awareness	degrees of knowledge
mind dormant	mind awake

(*Raj*) <u>active information</u> **is only created in conscious, choice-flexible mind.** *In the yellow illustration we saw what, for your microcosmic part, constitutes the order of your personal acts of creation.* <u>*Could this order of vertical causation reflect a macrocosmic, that is, universal pattern?*</u>

Idea is the seed. There accompanies intention to elaborate the embryonic inspiration's promise. This, grand or trivial in scope, is the basis of all purposes. Can we further understand the order of an act of creation from the way we seek to elaborate, that is, express an idea? From trivial creation of finger-wagging up to creation of a work of art or science all involves, essentially, the same order of process or, as already noted in that illustration, **phased intent**.

tam/ raj	Sat
expression	Idea/ Purpose
↓ tam	raj ↑
passive element	active element
informed	informant
physical elaboration	conceptual elaboration
created outcome	creativity
output	input
end-product	development inc. manufacture
hardware	software
machine	program

The outcome may be simple (as in finger-wagging) or complex. Complexity itself is of two kinds.

The first, <u>purposive complexity</u>, is psychological. *It is a function of information gain, expansion of consciousness and a capacity to grasp principles and purposely, creatively exploit the principles and possibilities inherent in any circumstance.* Conscious creativity is filled

34

by information, will-power and desire. *Indeed, will and desire are the psychological equivalent of physical electromagnetic radiation.* Will is like electricity, desire like magnetism. Volitio-attractive; pushy will and pulling of desire. These are the complementary energies of mind.

An increasingly concentrated focus of attention wakens to greater capacity, flexibility and possibilities for specifically ordered, coherent or *active complexity*. *Such purposeful complexity* works against the 'downward' wear and tear of time and chance; it codifies and specifies design - which chance cannot. It is an instrument of biological survival, intellectual enquiry, technology and artistic creation. We continually experience it. Its proof, in artefacts and actions, pervades our lives.

Transmission of purpose involves, in the outward direction, imagining and planning its realisation; and, on the inward, deciphering another's plan. This is active teleology. In fact, whenever an intention is physically expressed it works through mechanisms and machines. Machines, which include biological bodies, operate according to physical law and fulfil their function using, in one form or another, energy. They specifically accord with plan and inform the world in ways unguided nature can't. As such they obviously link information (that is metaphysical) with energy (that's not).

Such causal reason generates meaning. *All active information is meaningful.* Implementation of a plan involves *semantics whose specific meaning transcends the generality of grammar and syntax.* The latter are simply vehicles of reasonable expression; their immaterial symbols are needed to make connection with the material world and thereby order it. They translate mind to matter. Coding and decoding, using speech-through-air, written word or other forms of signal, are core semantic business.[24]

Information's Infrastructure - Code

↓ irrationality	logic ↑
no-message	message
no-code	code
physico-chemical maelstrom	psychological scheme
accident	teleology/ purpose
chance construction	technology
mindlessness	mindfulness

They make sense. Although *meaning* is the important, active ingredient behind codes, data transmission and storage, nevertheless these passive instruments of communication are important. We need frameworks (hardware) within which to manage information (software).

A command erases randomness; it is a deliberate restriction imposed to cause a non-random outcome. It employs an agent of restriction - sign, symbol, code of one sort or another. *This is the world of signals, semantics. What is spoken is not a matter of chance.* **Whatever is encoded is intentional**. Code and incoherent chance are chalk and cheese. They never mix. Language, other symbols and the construction or decipherment of meaningful communication (called semantics) make up the third level of information. In other words, this is the level at which ideas are framed in code or blueprint before their presentation in material form. *It is important when considering biology, whose basis - metaphysical basis - is code.*

On the other hand, *(tam)* passive information **is the expression, external to conscious, flexible mind, of active information.** Such information may be dynamically exchanged (as in the case of speech or body language). Otherwise its impressions are stored either in subconscious mind (featuring relatively inflexible memory, instinct and so on) or using matter (where instruction is carried by arrangement on chemicals such as clay, papyrus, ink, *DNA* or other messengers). Storage may be fixed (as in a file or photograph) or dynamic (as in a running film, program or automated mechanism).

Note that the second, non-purposive kind of complexity is the product of reflex forces and particles; oblivious, physical reactivity creates aggregations such as stars, mountains or snowflakes. Such complication is mindless, aimless, *passive*, purposeless; and its universe could not create a cup of sweet tea in a billion years.

To cut the story short, *the upper level of code is its abstract infrastructure, the lower is its physical arrangement (in chemical or energetic e.g. vocal form). Metaphysical precedes physical.*

Top-down, therefore, chance loses out. Code is always the result of a mental process. If a basic code is found in any system, you might conclude that the system originated from a mental concept, not from chance. You might therefore conclude it had an intelligent source - especially if that code is optimised according to such criteria as ease and accuracy of transmission, maximum storage density and efficiency of carriage (such as electrical, chemical, magnetic, olfactory, on paper, on tape, broadcast etc.) to its recipient; and if, above all, it works and orderly instructions are unerringly responded to.

A coder takes no chance. Randomness is eliminated. Compiler and processor are mindless mechanisms but a programmer is not. He determines the code and its operation: error is rigorously debugged. By definition, mistake or randomness degrades information; and the job of any editor is to eliminate interference, 'noise' or mistake. As we see, chance neither creates nor transmits information. On the contrary, accidents always (unless accidentally reversing a previous degradation) degrade meaning and, by degree, render information unintelligible.

Moreover, if information is immaterial then a frame (such as the holistic one that I propose) is needed; the metaphysical needs be included along with the shock from subsequent theories of Source, Origin or Potential to materialistic physics, psychology and biology.

All active information (whose psychology we'll address later) is meaningful. Implementation of a plan involves *semantics whose specific meaning transcends the generality of grammar and syntax.* We need frameworks within which to manage flows of data.

Code is devised and stored in mind but information's physical expression obviously employs material arrangements. The alphabets of such code include, as well as humanly devised systems of communication, bio-chemicals (supremely, *DNA*), binary nervous code and the sub-atomic elements, forces and atoms of physics and chemistry. The former couple are specific to life forms - possibly, at least with respect to *DNA*, anywhere life may exist in cosmos. The latter also constitute a universal code. This code dictates, through the agents that a study of phenomena elucidates, the way things naturally turn out. (*Looked at this way cosmos is a Grand, Dynamic and Encoded Text*).

The lowest level has, therefore, already been explained. This is the part that bottoms out in light, sound, fluid patterns and in crystalline solidity. This is the 'external', 'objective' or physical side of the psychosomatic border; and, while forces and atomic particles comprise its primary expression, the hard, bulk universe that we survey is its secondary. *At this level we find data items, that is, materials whose arrangement is governed.*

Finally, therefore, let's examine three lowest level but still complex expressions of purpose. Verbal, musical and mathematical codes have produced the glories of humanity.

Music.

The old word for integrated order is harmony.[25] **Music, like health, is harmony in action**. It is an archetypal formulation of energy, constrained only by its type of instrument, harmonics and the skill of its musician. Melody is a most profound form of information-in-motion. It is perhaps the best medium for the vibratory transmission of meaning. Thus, regarding composition, First Cause does not need brain!

Logical Archetype, primordial vibration's song, will do! Symphony composed by resonant association is enough! The vibratory energy of First Cause and its subsequent constructions would embody the internal logic, patterns or rhythms that pulse through each level of the grid of creation. Therefore, beyond the other cosmic models and although it can't be posted in a diagram, music is Natural Dialectic's Master Model.

Machine.[26]

The table you are sitting at is a device of simple and unmoving parts. Its construction serves a purpose and, therefore, has 'mind in it'. Its shape is 'passively informed' from chemicals that would never take that shape without direction. In fact, every organism marks the world with trivial, crude or intricate constructions for survival and, for some, passion's further purposes. Look around. This room is everywhere imbued with mind. Physical and chemical analysis, howsoever exhaustive, will never of itself find mind and, therefore, reason in its material parts. The hidden factor in plain sight is information. Everywhere you easily detect informants and informed. Why, therefore, is information 'outside science' just because it's metaphysical not physical?

"Machines", argued philosopher/ scientist Michael Polyani, "are irreducible to physics and chemistry." They are irreducible because they involve immaterial purpose, the stepwise development of a plan of implementation, a directed cohesion of working parts and, of course, the thoroughly non material anticipation of an operational outcome.

Let us rephrase. *A machine involves a system of well-matching, interacting parts that, unless any is removed or degraded, contribute to a function or produce a targeted result. Such systems are therefore specifically and irreducibly complex; <u>and so to work they must be made at once.</u>* In this case which comes first - a machine or its concept, a work of art or its inspiration, the chicken or its egg?

<u>Let us finally be absolutely clear that, although machines are operationally subject to the constraints of natural forces and environmental context, they are never created by them.</u> *No machine ever appeared as a consequence of the addition of random free energy into an existing system.* **Physical nature can't create machines; yet information and machinery are closely twined in living and, indeed, all teleological constructions.** *You want evidence for holism? This is fact.*

Mind Machine.

Mind is a symbolic representer and recorder. A computer is a mind machine. Inspect the logic of its functionality. Examine integrated circuits. Their molecules, like those of a brain, show no sign at all of mind. They are not even biochemical but metal, plastic, silicon and so on. Yet they are replete with passive, rigidified order. In this respect each one's determination cries out its ghost, the active order of its maker. The whole machine is absolutely full of maker's information.

Programs, coded programs **are a mind machine's intent. They express the will that mind invested in machine.** *It needs be re-emphasized - <u>mind</u> seeds the development of machine. Every machine (including body biological) expresses the ingenuity of its maker.* Not in its atoms *per se* but in its purpose, design and lawful operation. Machines passively embody information. Mind-machine information is passive. Active has produced passive information. *There is, to conclude, known neither law nor process of nature by which matter originates information.* Information is the immaterial element.

Chapter 3:   Psychology

Bear in mind, as we approach psychology, that Natural Dialectical Philosophy *is* internally self-consistent. Let's equally and importantly remember that, as in the finger-waggle, its knock-on networks of causation are not simply 'horizontal' (as with one-tiered materialism) but its 3-dimensional structure includes 'vertical' causation.

Lao Tse is seen here with Swiss psychologist Carl Jung. **Tse's cosmic fundamentals and the basic form of Natural Dialectic (the dynamic interplay of complementary opposites) are illustrated below.** Jung was deeply influenced by the Taoist I Ching or Book of Changes.

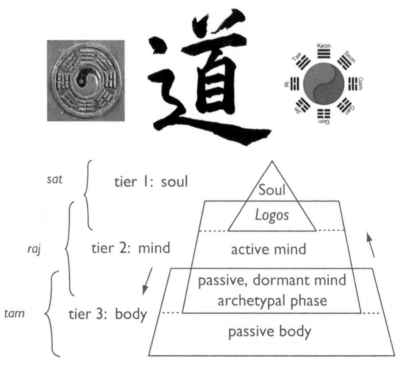

tier 3/ tier 2	Tier 1
passive/ active information	*Potential Information*
body/ mind	Psyche/ Soul
lesser selves	Transcendent Self
material/ mental forms	Conciousness
↓ physic	metaphysic ↑
energetic	informative
passive information	*active information*
non-conscious	states/ grades of mind
body/ tier 3	mind/ tier 2

With the idea of a 3-tiered universe do you also recall the logic of a contemplative monk or yogi - man microcosm of the macrocosm?

In meditation they seek to remove the existential motions that arise in body (matter) and in mind. Once the waves of mind are tranquillised they no longer obscure *Psyche* (Transcendent, Pre-active Truth).

motion	Essence
finity	Infinity
relativity	Absolution
appearance	Truth

This Informant Subject[27] forever precedes the forms it predicates, forms composed of you and every other thing. Infinite projects the finite. Essence both transcends and yet composes its existence, that is, informs its programs and substantiates the energetic patterns of creation. This conditionless condition is, therefore, Supremely Paradoxical. It is the Being of Potential Information communion with which was, in the previous chapter, identified as every mystic's goal.

> *Such Supreme Paradox is, beyond existence,*
> *Absolutely Non-Sensible; absurd as any quantum*
> *trait it is, however, furthest from absurd; it is our*
> *foundation and our starting-point.*

First Motion, first restriction of the Infinite is called First Cause, Causal Mind or Super-Conscious Archetype.[28] The previous chapter noted many labels; here the ziggurat locates it as *Logos*. It also noted that, as well as primary metaphysical, there is also secondary physical first cause. Upper and lower initiators. The latter, unconscious archetypal phase (tier 3) features in the previous, this and the next two chapters.

However, this is all *materialistic nonsense.* Claptrap.[29] So you have never met such notions in the context of a scientific course. **Materialism deems that thought and subjective experience (by sense or otherwise) are things**. But could senseless atoms ever yield their own experience? Or mathematics? Reasoning? Morality? Intelligence?

Remember that the world, for a materialist, is basically colliding particles and interactions due to force. *And, biologically, neither cells nor their nucleic acids know or care a fig.* Such oblivion is ignorant of laws of logic, creativity or purpose. Why should compound intricacies of atoms, up against the laws of entropy, codify for algorithms and the mechanisms for an intricate survival? Even *if* blind carelessness developed nerves (a great guess in doctrinal dark) why should their chemical reactions produce a calculation or a sonnet? Physicist **Sir John Polkinghorne** rams the point home. '**Thought is replaced by electro-chemical neural events. Two such events cannot confront each other in rational discourse...The very assertions of the reductionist himself are nothing but blips in the neural network of his brain. The world of rational discourse dissolves into the absurd chatter of firing synapses.**'

Please, therefore, allow at least the axiom that information, active or passive, might exist in immaterial form; that mind is metaphysical.

The word *'psychology'* means study of *psyche* (from a Greek word generally translated 'soul').[30] But current 'psychology' were better termed *'noology'*. This is because the discipline studies mind, for which the Greek is *noos* (from the verb 'to notice' or 'to think'), and not *psyche*. Or does it? Perhaps not even 'noology' goes far enough. **Scientific materialism can, by definition, only allow physical composition.**

Upper, Psychological Sub-divisions of the Ziggurat

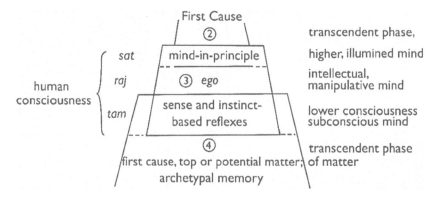

42

So what does the word 'soul' or 'psyche' mean? Inner essence? Centre? Does it include immaterial consciousness? *Bottom-up*, it is figment of brain's motion, an imagination born of nerves.

Top-down, however, might you not reflect creation as a whole? Is your own constitution, microcosm of the macrocosm, in the image of Chapter 2's three-tiered universe?

Having earlier registered qualities of *Psyche* (Soul) and *Logos* (First Cause) it is at this point noted that, from a *top-down* point of view, created 'mind-with-body' is seen as a 'soul vessel'. Thus every different pot (or organism) contains the same pure water. In this manner a human may be seen as soul having a physical experience rather than a body erroneously imagining its metaphysical soul. Also, in this view, who masters best his subject is best qualified to treat - top-class and wise soul-doctors (called psychiatrists) are saints.

Clearly, we have two very different views of mind and soul.

Substitution, metaphysical for physical, is, however, scientifically 'unacceptable'. 'Impossible' - whatever future studies might disclose. 'Immaterial' is a word too far. *So, naturalistic creed decrees, thought's entirely a nervous matter.* **Thus, naturally, psychology emerges as the study of neurological phenomena**. Carbon, oxygen and more, from this perspective soul is the activity (incredibly, incomprehensibly complex, mind you) of particles. Consciousness 'emerges' from non-conscious molecules grouped in some special way. Isn't thinking generated by the soft-wired workings of a brain? *Just sling sufficient atoms, in the form of nerves, together - they'll become no less than self-aware!*

Holistically, however, that mind is a physical illusion is the neurological delusion.[31] **To mistake its neural correlate (as, say, registered in brain scans) for experience itself is an error as basic as taking the electronic pulses in a wire for all there is to telephonic conversation. It is a prime, elementary fault, a first category philosophical mistake; it might be termed full-blown, psychological mythology**. And to identify consciousness as a material illusion is itself, denying the reality of one's own experience, a pernicious - even dangerous - delusion.

One party, it is clear, *believes life is a phantom of the atoms*. Brain *causes* subjectivity. Thought (therefore belief and all the purposive effects of hope and will) is part and parcel of nerve chemistry. And what is the *experience* of consciousness? The essentially robotic view of neuroscience holds that nerves *are* consciousness. We just don't yet understand, the faithful purr, how brain's 'emergent properties' can squeeze experience out or how the juice that's 'you' must be exuded from its molecules. A revelation is, however, prophesied. Materialism's scientific certainty decrees that life will be reduced to chemistry and mind experimentally identified as simply due to complicated ionicity! You are a product of your physiology and so, at root, your genes alone. Life has, hasn't it, to be an electronic after-thought?

Nervous particles and atoms aren't, like atoms anywhere, alive. Therefore, if life is made of them it shouldn't be alive! *Thus the other line suggests, conversely, that brain isn't an originator but a mediator. It is our dashboard as we fly through life.* A filter. A sophisticated interlink between mind, body and the latter's physical world.

If so, it is a transducer device that, like any mediation network (e.g. radio), must be sufficiently well-constructed to handle large volumes of two-way traffic. It accepts environmental signals and translates them (↑) 'upward' into mind's experience; and issues orders (↓) 'downward' into body chemistry. As an organ of 'cockpit control' its 'dashboard' accurately connects an immaterial mind to a material body and, thereby, physical conditions. In this view mind and brain, although compounded, are quite separate entities - the former metaphysical and latter physical. Brain chemistry's identified as a design that expedites exchange of information. Your head is thus a medium!

Making no material difference by adding immateriality, the Dialectic simply reconstructs creation on the basis of a 'conscio-material' duality. In short, perhaps brain neither does nor ever did enjoy a seamless, subjective experience. *The implications of this seminal idea are so extensive that this whole course is exploring them.*

So now to <u>*Consciousness*</u>.[32] This is what it's all about. Without it you are nothing. The star of every play is mind; the kingpin of psychology is consciousness. What is the 'thing'?

Non-conscious matter is a special case of its subjective absence. *It is pure non-consciousness.* Gases, streams and solid bodies don't know anything. Their oblivion's polar opposite is total wakefulness. Of what, you ask, does this consist? As matter's pure non-consciousness exists could not a concentrate of immaterial information - pure consciousness - have being too? So that creation's root turns out to be oblivion's antipode - *Potential Knowledge, Latent Field of Knowing, Pure Concentrate of Consciousness.*

Materialism doesn't like this sort of phraseology at all.

The stack at the start of the chapter illustrates <u>triplex psychology</u>. Its three phases are Super/ Ultra-conscious, conscious and sub/ infra-conscious.

These 'gears' in the system act like levels in a cosmic grid, a chain of command. This chain applies equally to our microcosmic selves. It maps as hierarchical psychology.

Top-down, if pure non-consciousness (matter) forms the cosmic sink its source is pure consciousness. *Psyche. Soul. Potential Information. Uncreated Consciousness is precedent and therefore primary; the lower, existential range of forms is secondary*. What light is to physics Essential Transcendence is to psychology; in other words, the only absolute measure of consciousness, and therefore psychology, is Transcendent Super-Consciousness. From mind holistic distillation therefore reaches high; but materialism recognises only base degree, the lifeless one.

Hierarchical Psychology: Mental Ziggurat

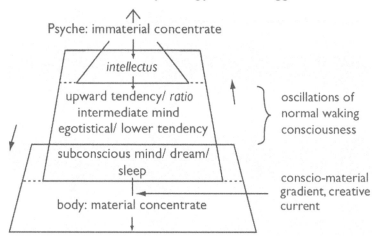

From *top-down* perspective, any attempt to reconcile physical science and psychology must logically start at the top rather than bottom extreme. It must start at Axis, Centre, Source or Pivot (Conscious Soul) rather than with the effects of subsequent informative or energetic motions. Thence we descend from the level of super-state transcendent (*psyche*) briefly to the place of our restricted, human awareness.[33] Next we fall to cover, at the sub-conscious end of mental balance, the conditions of dormant mind, dreaming and deep-sleep; finally we inspect the psychosomatic domain of instinct, personal memory and archetype. The latter account for the psychosomatic connection between mind and its sub-state, non-conscious material body.

Vectors and States by Stack

spectrum of consciousness (2),(3),(4),(5)	Super-Conscious (1)
↓ downward tendency	upward tendency ↑
passive/ exhausted	active/ alert
inertial equilibrium	dynamic equilibrium
non-conscious/ subconscious	conscious
busy-asleep/ dreaming theta (4)	busy awake beta (3)
deep sleep/ coma/ delta (5)	contemplative/ alpha (2)
involuntary	voluntary

We can divide these 'bands' or phases slightly further to relate five main states of mind with their correlated conditions of brain. Which 'gear' are you in right now?

Five Main States of Human Mind in Relation to Brain

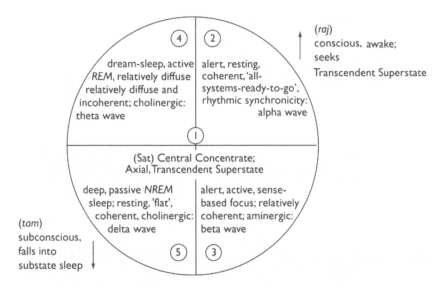

(4) (2)

dream-sleep, active | alert, resting,
REM, relatively diffuse | coherent, 'all-
relatively diffuse and | systems-ready-to-go',
incoherent; cholinergic: | rhythmic synchronicity:
theta wave | alpha wave

(*raj*)
conscious, awake;
seeks
Transcendent Superstate

(1)

(Sat) Central Concentrate;
Axial, Transcendent Superstate

deep, passive *NREM* | alert, active, sense-
sleep; resting, 'flat', | based focus; relatively
coherent, cholinergic: | coherent; aminergic:
delta wave | beta wave

(*tam*)
subconscious,
falls into
substate sleep

(5) (3)

Of these states, as also shown in the previous stack, 1 relates to Super-consciousness, 2 and 3 to waking and 4 and 5 to sub-consciousness.

Here is brain drawn simply.[34]

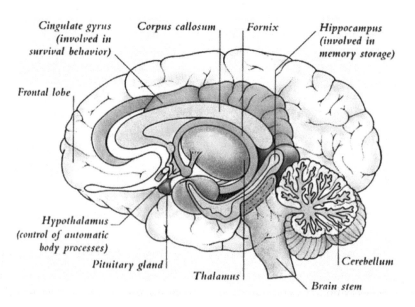

Cingulate gyrus
(involved in
survival behavior)

Corpus callosum

Fornix

Hippocampus
(involved in
memory storage)

Frontal lobe

Hypothalamus
(control of automatic
body processes)

Pituitary gland

Thalamus

Cerebellum

Brain stem

Libraries have been written concerning the state of psychological normalcy, our waking state (here 2 and 3). Suffice here to note that *ego* is a mask or per-sona. Its dynamic structure, necessary for function in a body, amounts to a structure, face or formful covering of inner, underlying consciousness. It is this reducing agent that a contemplative attempts to 'vaporise'.

We are now in a position to connect metaphysical with physical hierarchy, the latter as it appears in the construction of the most complex object in the universe - a human brain. Claimed to have evolved out of a total lack of logic it can, however, produce a logical and comprehensive understanding of itself. Boot-strap logic here, *par excellence*!

Bottom-up, brain is seen as the generator of thought. Nerve cells, made of non-conscious molecules, create the experience of an illusion known as you.

Top-down, on the other hand, brain is seen as a mediator, a complex dashboard, a fine linkage factor between metaphysical mind and physical body to which, for a limited period, it is attached. The brain is a machine, a mind machine like a computer, and this is its task.

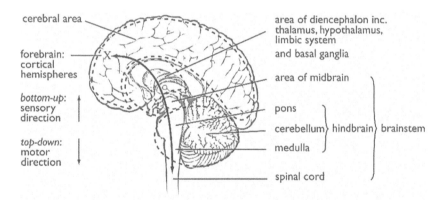

These dialectical drawings of the logical, *top-down* construction of the brain, are at the same time an organic representation of the levels of mind.

Was brain thought out? Is there, after all, no logic in the way a brain is built; or is it just an aggregate of happy, codifying accidents? *It is logically surprising that, for no reason, non-conscious and illogical chaos should have constructed order to a very highly systematic climax in the most complex working system of the cosmos, an information processor whose whole, sole, negentropic business is order - a central nervous system and associated brain.* Did matter, getting far more than it didn't bargain for, perchance 'evolve' a brain? Materialistically speaking, it must have. *In reality did it?*

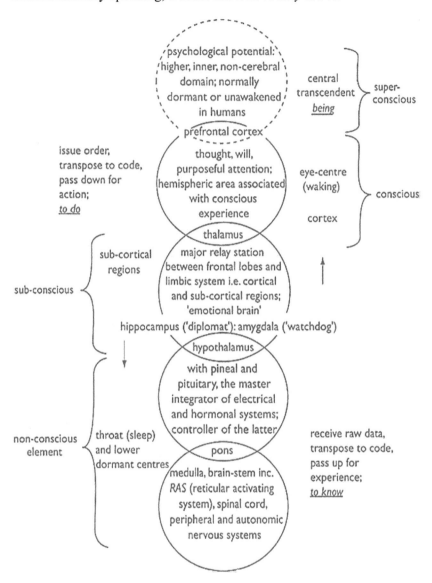

A brain so simple we could understand would be so simple that we couldn't; but, *top-down*, we can enumerate the principles round which its great complexity might gather.[35] Relate this illustration to the three preceding it. Does not the order of its parts confirm a *top-down*, hierarchical construction of our dedicated organ of intelligence?

From conscious mind we now descend to the subject of *sub-consciousness*.[36]

The Sub-conscious, Psychosomatic Sandwich

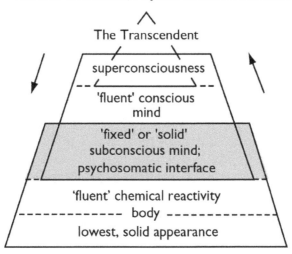

Are you ready for a fall to the diffuse conclusion of psychological entropy, for a subjective drop into the labyrinth of underworld? You know what it is to be mentally as well as physically exhausted. You've often dropped off into mind's flat, dark condition we might call inertial equilibrium. 'Little death' is not the world's end so let us take a snooze cruise; it is time to fall asleep.

Dreaming. For dreamers dreams are real enough; but the experience is untrammeled by either external events or the ability to reason. Waves wash equally on what is in their path; a torch shone randomly around picks out disconnected or illogically connected objects and events. The files are scattered, narrative is blurred.

Deep Sleep. In deep, non-*REM* (or *NREM*) sleep the 'upper', voluntary structures of brain are cut from the loop. A sleeper's movements, including eye movement, are much restricted; sensation is dull or absent. Brainwaves, the overall coordinators of the central nervous system, slow to between 0.6 and 3 hertz. These are so-called delta waves. Maybe deltas drop to zero. Brain death. If, by head injury, stroke, tumour or poison, the sleep/ wake toggles fail or signals cannot reach the forebrain then the patient drops into oblivion. The curtain falls but drama does not start again. Coma is an open tomb, an unpinned shroud or wake-less sleep - though in its stillness deeper grooves of mind (archetypal constructions but also profound personal impressions and rote such as language) stay frozen yet intact.

Frozen Time. You sleep but your past does not disappear. You wake and your past has not disappeared. You think you have forgotten, you may even suffer amnesia but untapped memories remain. They are how we freeze time. A memory is frozen time. It symbolically encodes the past. *A memory is a thought object and, as such, has no life of its own.* A disc encodes music once recorded 'live'; it's a memory that, when replayed, affects the present and, from this, the future. Thus mental memory is an encoded image; it is a record and, on conscious recollection, becomes a presence of past action that may affect the future.

Lower, Unconscious Sub-divisions of the Ziggurat

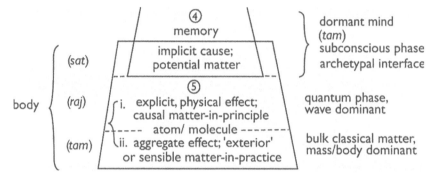

From a top-down perspective, **neither memory nor knowledge are inherently physical.** Mind is a wireless entity. No doubt, correlated nervous circuitry acts as a storage-and-retrieval system that, by association, allows the immaterial library of remembrance its efficient, selective interaction with an innervated body and its circumstances. Thus so-called 'engrams' (are they sited at synapses, inside a nerve cell's body or elsewhere?) may, if they exist, indeed *relate* to physical experience; they may act as a recognition trigger, reference point or body's resonance with an experience. And organs (such as hippocampus and amygdala) certainly seem to log experience in the manner of a record/ playback head; they catch or release a moment that, in fixity, is called a memory. But if such 'storage' or 'playback' button fails the system's compromised. Either records are not made in the first place or the connection becomes impaired or irretrievable. *But the 'disc' of memory itself is metaphysical.*

Memory, the only form of metaphysical information storage, is the shape of infra- or sub-consciousness. *Indeed, it is sub-consciousness.* Subconscious mind consists of coded files that we call memories. It is made of memories. Regarding life-forms memory comprises an organism's library of precondition and conditions, that is, its context for experience. The precondition is its archetype, the basis of its sort. We might call this sort of memory '*typical*' while '*personal*' experience includes both active (created and transmitted) and passive (received) information. Either kind of memory is 'frozen' like still photograph or film. It may operate like a movie or store programs that, once triggered, can unfold like stories in a sequence to their completion. Who has ever committed a poem to memory? Any mentally stored plan is such a memory. Programs are, although dynamic, still a frozen form of mind; and they're replete with information. They specify the most efficient means to a well-defined end. You might argue in this vein that biological structures are codification incarnate; and that the concept they express is an archetypal program.

Archetype[37] we see as a First Cause. There are First Causes, psychological and physical. **The latter's archetype (potential matter) will, along with quantum matter-in-principle and matter-in-practice,** (see Chapter 4: lower sub-divisions of the ziggurat) **lead us to a triplex view of the bodies of physics and biology.**

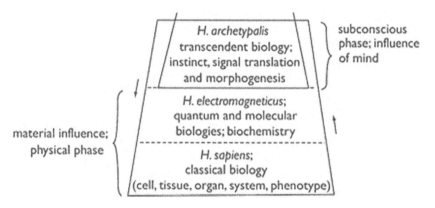

This diagram illustrates lower sub-divisions of the ziggurat in terms of internal hierarchy and the place of *H. archetypalis* in biology.

Archetype is, like any idea, the informative potential for its own later expression. The word means basic plan, informative element or conceptual template. Its type is, according to holistic logic, held in subconscious mind; archetypal memories compose the subconscious element of universal mind. Thus it consists of pattern in principle; it is the instrument of fundamental 'note' or primordial shape, the causative information in nature or 'law of form'. It is nature's *bauplan* or blueprint. **Called potential matter its fixed files are seen as hard a metaphysical reality as, say, particles are physical realities.**

Of modern psychologists Carl Jung had perceptive sympathy with the idea of archetype. Although he did not cast his net wide enough to include all organisms and the whole inert part of cosmos he nevertheless suggested that human archetype involves elements of the 'collective unconscious' or 'universal psychic structures'. Such 'psychophysical patterns' are given various individual and cultural expressions by human consciousness. Furthermore Jung uses the analogy of light's visible spectrum wherein normal human consciousness is yellow with unconsciousness at both red and blue ends: Natural Dialectic's conscio-material gradient likens the visible spectrum to normal mind with ultra-violet scaling super-consciousness and subconscious infra-red grading down to extra-low frequencies of non-conscious matter. Thus Jung also sees the archetype as a bridge from mind to matter. Finally, crucially, just as local objects and events are details distinct from the general principles that govern them, he differentiates between specific, local, conscious imaginations (or 'motifs') and the 'unconscious' generality of archetype from which they derive.

Not only Carl Jung but founding fathers of neurophysiology addressed the mind-matter (psychosomatic) issue in which archetype plays, dialectically, such a pivotal role.

"What greater inherent improbability", wrote Sir Charles Sherrington, "than that our being should rest on two fundamental elements than on one alone?"

In this case, how does mind interact, primarily at least through brain, with body? What might constitute the nature of an interface (*PSI*, see Glossary)? We turn to the bio-classification of psychosomatic linkage.[38]

Three diagrams help illustrate a suggested architecture of the subconscious[39] for both conscious and unconscious organisms. **And as every cell contains chemical genetic information, so it is suggested each also acts as an antenna for its typical mnemone, that is, its archetypal broadcast.** In other words, as well as by 'holographic' *DNA*, each cell is informed by subconscious, archetypal memory. Every cell of every organism involves mind.

Firstly, your dialectical bio-classification.

Grades of Man/ Dialectical Bioclassification

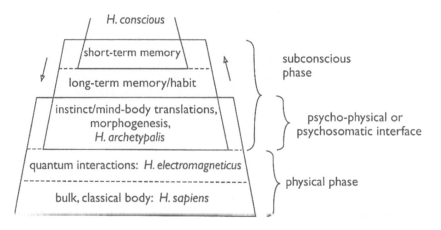

Secondly, your bio-classification corresponds with the rings of Wireless Man. Its typical mnemone (*H. archetypalis* or 'memory man') is, in effect, a channel carrying the program of human instinct. Positioned at the mind's side of the ring representing psychosomatic interface, its physical correspondent is *H. electromagneticus*.

Wireless Man

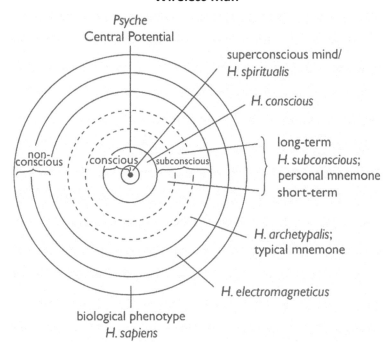

Psychosomatic Linkage by Domain

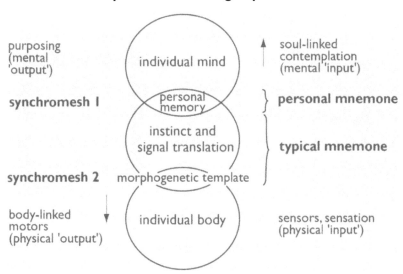

Thirdly, as shown in Psychosomatic Linkage by Domain, typical mnemone is composed of three main sub-routines; or, if a protocol is a standard procedure for regulating the transmission of data between two end-points, three linked protocols. Together translational, instinctive and morphogenetic programs comprise an individual's archetype. More detailed architecture can be found elsewhere.[40] If, as in the case of plants, fungi etc., there is no conscious component, then the translator element is reduced from nervous to chemical (e.g. hormonal) messaging alone.

The basic principles of archetypal psychosomasis are by now clear. *Mind (at its most gross, subconscious level) conjoins with matter (at subtlest, least-massive/ almost-immaterial level); elusive quantum probabilities pinned-down substantiate, it seems, certain processes; photon, electron and their effect on quark (protonic) or atomic position precede, in the sense of govern, molecular and bulk reactivities; and, where electrodynamics describes the effect of moving electric charges and their interaction with electric and magnetic fields, biological electrodynamics precedes all bio-molecular considerations.* **Every biological process is electrical; and the flow of endogenous currents is the primary and not secondary feature of physical life. Not only biochemistry but quantum biochemistry heave to the fore. Natural Dialectic lifts perspective from molecular to a vibratory, field perspective.** *It is thus suggested that, at electrical and wireless levels, patterns of subconscious mind meet and influence matter; archetypal information is relayed to chemistry by polar charge and light.*

This subtle, wireless body (*H. electromagneticus*[41]) has, in the human case, been identified as the physical side of message-exchange and *H. archetypalis*, with its routines, the metaphysical calibrator.

There is no doubt, the universe is in vibration. The cosmic transmission of information and energy is, at root and as spectroscopy confirms, vibratory.

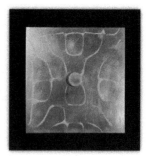

Ordered oscillation is called harmony; harmony is the grammar of music and music is a universal language. The theory of music is implicit in any recital. Could it transpire that explicit order of a cosmic recital is the product of implicit notation? Could its performance represent cooperative forces, specific 'notes' called particles and thereby, all in all, harmonic code?

Here we note the work of, among others, Ernst Chladni (above) and Hans Jenny. **Their study, now called cymatics, involves the use of wave frequencies to precisely control the production of shapes in water, air and other media.** It is significant that Chladni's figures often imitate familiar organic patterns that we see in nature and, especially, biological structures.[42]

 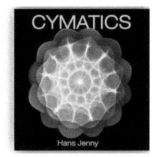

Pattern clearly relates to frequency of cycle. And resonance, whose orderly aspect is characterised by an analogy with music, is the tendency of a body or system to oscillate with larger amplitude when disturbed by the same frequencies as its own natural ones. *Cymatics therefore intimately involves the vibratory transfer of energy.* Such transfer is an integral part of all vibratory systems wherein waves interfere with/ destroy or cohere/ amplify each other. Energy is amplified and transferred by resonance and attunement. Common examples of electromagnetic resonance include tuning a transmitter/ receiver and photo-electric initiation of the photosynthetic process, that is, the first step in life's chemistry.

For quantum physics matter is certain vibratory frequencies of energy; and, from a dialectical point of view, it is simply stresses, strains or tensions in the medium of their absence, that is, nothingness. **There is, however, nothing random in a highly orderly creation derived from first acoustic principles. Oscillation between polarities, cycles, vibratory rhythms, <u>the interrelationship of waveforms and complementary resonance are at the heart of Natural Dialectic</u>.**

And it is suggested that a key phrase in the suggested explanation for the wireless, psychosomatic transfer of information between subconscious structures of the mind and the physical plane is '<u>resonant association</u>'. *The modus operandi of psychosomatic broadcast is therefore, in a word, attunement. Resonance.* It involves entrainment between typical mnemone and *H. electromagneticus* or, if you like, transduction between recorded information (memory) and physical energy.

Along with instinct and signal translation the *third* component of a typical mnemone is the **morphogene**. Of course, if you equate morphogenesis with molecular biology alone then 'memory man' is superfluous nonsense. Materialism ridicules the idea of natural informative structure, a metaphysical 'cloud' or mnemonic database called archetype You might well, if morphogenetic *vis medicatrix* is other than chemical, deny its clinical influence. What cannot be physically explored does not medically exist.

No doubt, 'solenoidal' *DNA*, hormones and other codified biochemicals are intimately linked with growth, development and morphogenesis but are they all there is? And do we suppose these microsystems or a cell's or body's perfect health evolved through grades of imperfection, that is, millions of sicker stages? Or prefer that archetypal vigour radiates original, dynamically fine-tuned health for every kind? *If your body's health is indeed the result of interaction between physical and archetypal (sub-conscious) patterns, it did not evolve by accident.* The tendency of every system is to bounce back to its archetypal norm, its *vis medicatrix*.

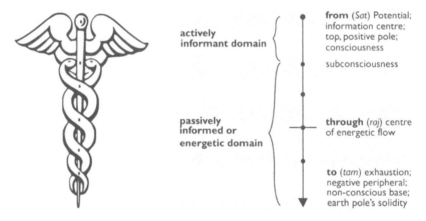

This metaphysical form is the very basis of homeostasis, physical life's overriding principle. Health is co-orderly; dynamic equilibrium is the norm. Wounds heal, infections are fought and cell debris cleared. Indeed, the central nervous and autonomic, endocrine and immune systems are integrated in such a holistic way that experts sometimes use the phrase 'psycho-neuro-immunological system' to describe their cohesion. Mind's attitudes and purposes substantially affect, as every doctor knows, the body. The channels of 'psycho-somatic interaction' are precisely what this chapter is describing - unless you still neuro-scientifically believe that mind is brain and psychological equates, essentially, with chemical.

Not only 'memory' but 'information' man.[43] The above-pictured wingèd staff, ancient symbol of Hermes, remains to this day the totem of many scientific medicine men. From brain to base of spine it represents the human extent. In the way of Natural Dialectic, its

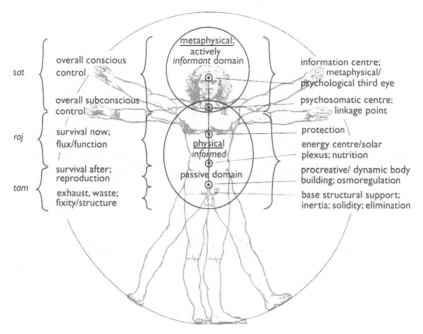

gradient reflects the linkage between informative and energetic domains of man. The microcosmic slope runs top to tail - from informative head section through an energetic centre (heart, lungs, digestive tract) to base systems of exhaust; from informant to informed by way of double helix it illustrates the logic of embodiment.

Here, as in all biology, reason's nowhere absent but, with chance, like chalk and cheese. *Why should a fundamentally rational system irrationally 'self-organise'? Your own human microcosm underlines the utter feebleness of evolutionary explanation.* If materialism's rational you could argue that it spotlights how irrational, backing chance as its creator, this species of rationality's become.

Isn't it, on the holistic hand, inconceivable that such a logical, integrated, self-consistent embodiment as yours, constructed with highly specific complexity in accordance with grades of the conscio-material gradient of creation itself, occurred by chance? **If reason wins, whose archetypal program is worked out in every bio-form, then chance and time's grand theory crumble back to their home ground - they bite their progenitor, the dust.**

To summarise the section on sub-consciousness, mind is linked to matter by a wireless anatomy whose instrument is the quality of vibration called resonant association. Our human song is intricate, a subtle system well worked and working out. Its own recital play in every other organism too.

Chapter 4: Physics

Friend Lao Tse now with Niels Bohr (the quantum theorist who, having showed that the electron orbits of an atom explained its chemical properties, neatly linked physics and chemistry) remind us that Dialectical foundations have stood the test of time. Bohr incorporated the dualistic, dialectical *yin-yang* design into his family crest!

Other modern seekers after physical truth include:

Note three points:

1.   Man is an information-craver. He wants to know more and more about his phenomenal surroundings and the working of his own mind. He wants, moreover, to find patterns by which to

order his knowledge to reflect as closely as possible the way in which his world works. Check boxes 1 to 3 of Chapter 1.

> **Physics, however, is restricted to answering questions purely with respect to and in terms of the non-conscious, reflex, physical world in which we find ourselves.**

2. Man wants to know it all and often, like a child, thinks he does. However, physicists are well aware how little we actually know. Although in 1900 it was thought that physics was, essentially, complete 2000 radically disagrees. Apparently we have only studied 5% of the universe, luminous matter. Current areas of attention with unsolved problems include:

   General physics/quantum physics

   Cosmology and general relativity

   Quantum gravity

   High-energy physics/particle physics

   Astronomy and astrophysics

   Nuclear physics

   Atomic, molecular and optical physics

   Condensed matter physics

   Plasma physics

3. Mankind may eventually completely understand the cosmos he lives in but, given the above, don't expect answers to every question in an hour or two! This talk fits physics, as far as we understand it, to the frame of ND (Natural Dialectic). It gives fresh context and perspective. This especially applies because material science cannot reach actively informative, metaphysical but still natural elements; and it cannot include this immaterial 'missing factor' in its measurements. Such factor applies to matters psychological but matters not one jot for accurate physical appreciation of cosmos; but since abductive logic and interpretation are required to deal with historical events, especially original and consequent order, regarding cosmos and biological (mind-matter) forms it may fall short. **Perhaps, indeed, human mind is not built to understand the beginnings but, being ever-curious, we'll work towards our best shot, that is, our holy grails.**

What, dialectically, are holy grails about? They are about achieving, at the heart of myriad, detailed complexity, simplicity and union. They are of two kinds - material and immaterial.

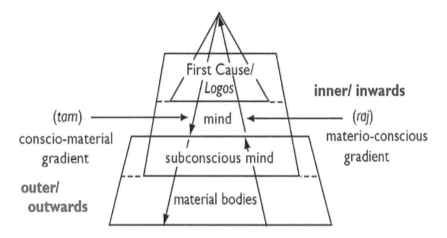

Do you remember this key illustration from Chapter 2? Scientists of physical phenomena and scientists of the soul travel in *opposite* directions. In spite of this both seek a Source and, as such, First Cause; but their perspective differs and their goals, immaterial and material, are actually antipodal. This does not mean both are not, in their way, right or that a man cannot be both scientist and a mystic.

The meditative focus seeks travel *up* a materio-conscious gradient, away from physic towards metaphysic, towards the source of mind and 'inside information'. Such grail is called enlightenment, that is, unification in First Cause. This Cause has, as we saw in Chapter 2, many names. The *Logos* is the *Tao* is The Informant. **It the Most Natural Heart of Nature whereas physical archetype is the natural heart of non-conscious physic.** In as much as Natural Dialectic provides a vehicle within which the causes and grails of physics, psychology and biology may be coherently, co-understood it may be called a Grand Unified Theory. Including both mind and matter it is, in effect, the *GUT* of *GUTs*.[44]

subsequent hierarchy	Holy Grail
issue/ action	Source/ Priority
physic	Metaphysic
phenomena	Causal Noumenon
classical/ quantum	Archetypal
↓ tam	raj ↑
outward finality	inner support
exhaustion	action
aggregate/ precipitate	influential action
large-scale body	microscopic components
matter-in-practice	matter-in-principle

Scientists concentrate, as we mostly do, *down* the conscio-material gradient; seeking 'outside information' they follow out through the senses and their technological extensions to learn the details of creation's unconscious shell, the material universe - including our own physical bodies. They seek, in a quest to find the **Holy Grail of Physics**, unification. **This, unification (by *GUT* or *TOE* - a 'theory of everything'), is the goal with its subsequent hierarchical organisation; but the maths of microscopic matter-in-principle (quantum theory) and large-scale matter-in-practice (relativity) do not fit together.** A joiner might be **quantum gravity** and theoretical attempts to create a union are taking place. Nor are dark matter and lambda force (anti-gravity) much in the picture. It seems a Higgs boson may, at very high energies, exist but it is more hoped than clear how this tiny creature produced that most basic of physical quantities - mass - and, if not, how did mass appear? Nor, of course, has any immaterial factor ever been considered. Universal mind is not considered. But now, in the attempt to integrate mind with matter, I want to investigate this stack of Holy Grail. Its three parts (metaphysic, matter-in-principle and matter-in-practice) are those of the lower three sub-divisions of the following ziggurat (first met in Chapter 3: Subconsciousness).

Below are a couple of alternative versions of lower, unconscious subdivisions of the ziggurat met in the previous chapter. These models is interesting because they introduce the notion of a Triplex Physics, that is, three stages of physics according to the cosmic fundamentals (*sat*) informative potential, (*raj*) action-in-principle and (*tam*) end-product, action in finished practice. **In other words, they prompt towards a Theory of Physic and Metaphysic.**

Everyone understands a ***horizontal, age-related chain of cause and knock-on effect***. We'll discuss such a first causative hit, perhaps big bang, a little later; but for now, reading *top-down*, recall from chapter 2 the causal cones. These illustrate higher and lower first causes in ***the vertical, informative sense.*** This pair can also be expressed in terms of Mount Universe.

Primary (Metaphysical) and Secondary (Physical) First Causes

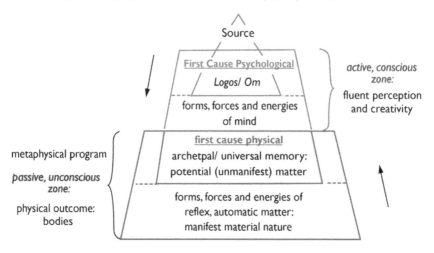

Archetype (see also Glossary) **means basic plan, informative element or conceptual template**. An archetype is the informative potential for a creation; as such it amounts to precondition. It is *first cause* that precedes action.

The two first causes are both, as we saw in Chapter 2, informative; but the higher psychological is active, conscious and creative and the lower passive, unconscious and fixed, as various moulds, stencils or files in archetypal memory. This latter, first cause physical, is the 'father' of physics, the generator of precise quantum vibrancies or forms. As thought is father to the deed or plan is prior to ordered action, so archetypes precede material phenomena.

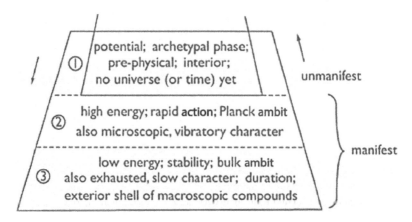

In the previous couple of illustrations the red-brown elements of ziggurat show the order of non-consciously energetic creation (reflected in eras of big bang theory[45]). Things develop orderly, as you might dialectically expect, from as-yet-unexpressed potential or archetypal level.

In short, archetypes comprise the *potential* phase of both psychological and physical triplex. **Archetypal source is therefore an important feature of holistic cosmos, the level whence things develop orderly.** Do you remember, from Lecture 2, thinking of it as a 'seed' which precedes the origin of physicality? In this case you might expect its metaphysical operation to follow different rules from those in physical space and time. And a seed is related to every part of the body developed from it. Could inmost archetypal seed exist, like a kind of holographic '*DNA*' within the body of our starry universe, apparently nowhere - being metaphysical - but actually everywhere and every time at once? If you think 'archetypal first cause physical' is a 'cop-out' in explaining how cosmogony proceeds then I suppose a systems analyst must think conceptual plans for any working mechanism or his working chips are 'cop-outs' too. However, philosophical objection to a program's purpose in no way mitigates the impact of its natural possibility and, if an immaterial element of information exists, it's natural fact.

We now turn to deal with triplex physics as regards its fundamental levels. We start with archetype - potential energy. There follow active matter-in-principle, the staple of quantum physics; and passive matter-in-practice, the bulk aggregates we call our sensible universe.

(Sat) Potential Energy

Potential precedes possible action. It is a prerequisite or precondition for results; *but Natural Dialectic does not think of physical potential in the way that physics does (e.g. potential gravitational or electrical energy).* **By now we understand potential energy/ matter is a metaphysical informant.** It is super-matter whose synonym is archetype. **Archetype is, like any idea, the informative potential for its own later expression. The word means basic plan, informative element or conceptual template**. Like instinct it abides, according to holistic logic, as a *form* or *file* of memory in dormant, universal mind. It serves as physic's data bank.

Thus it consists of pattern in principle; it is the instrument of fundamental 'note' or primordial shape, the causative information in nature or 'law of form'. It is nature's *bauplan* or blueprint. **Called potential matter archetypal files are seen as hard a metaphysical reality as, say, particles are physical realities**.

In the sense of *precursor* to expression of a physical reality archetype may be seen as a First Cause. *Check the lower subdivisions of the cosmic ziggurat and, from implicit cause through the explicit* *effects of matter-in-principle to the matter-in-practice of bulk or compound bodies, derive the notion of a **triplex physics**.*

With or without such cosmic program, we can reasonably claim invariance under transformation. Underneath the world's commotion character of basic parts remains the same. Contexts change but automated rules of play do not. *And without conserved invariance you can't obtain the balance that equations need.* **Energy's conserved**; and so is symmetry. **At any time in any space from any angle, physicists agree, laws of physics stay the same.** Does physics see such changeless cosmos as a self-consistent, fine-tuned set of principles or are its invariant patterns of behaviour basically informed by chance?[46]

Rules are invisible but they precede a game. Rules are information, information is potential for behavioural patterns. Regulation thus precedes and guides the way a game is played spontaneously.

In this case we say that in the beginning was *NOT* chaos. Modern physics shows that mankind dwells in a finely-tuned universe. To be precisely fit for life it must have started in a way most orderly, specific, specially defined. *Fine-tuned by chance? If low probability together with specific definition indicate design, no chance!* **A universe fine-tuned regarding many parts might be construed as one of specific, irreducible complexity; and, if lacking any part it failed to work, of minimal functionality.**

No objects bear a number yet you number them and count. Numbers, symbols and mathematics aren't a physical but changeless, metaphysical reality. ***Maths, while we're on the subject, is a real form of metaphysic.*** Pythagoras, at least, believed that 'All is Numbers'. Could essential, physically-independent numbers really govern physical complexity? Einstein, Planck and Eddington believed that, once you dropped upon their key, you could *deduce* (*top-down*) the reasons for all natural laws; you might unlock the codes whose inmost mysteries reveal just how a stable universe is sparked; a feat of mind *par excellence*!

For others, including Roger Penrose, the Platonic world of mathematical forms is also real. He writes, '*There is a very remarkable depth, subtlety and mathematical fruitfulness in the concepts that lie latent within physical processes.*'

Could such immaterials help describe mind-matter frontiers whose formations are called archetypes? **Archetypes are also metaphysical and have no being but in mind. Maybe 'natural mathematics' is describing real, archetypal files.** *Perhaps archetypes compose the link between the corners of what Roger Penrose has referred to as a mysterious triangle of physics, maths and mind.* At any rate, their logic's fixed. Fixed forms of mind are memories; thus they are memories in universal mind.

Could *you* plan a cosmos better? Lucid physicists agree it looks 'as if' designed. **When it comes to astrophysics and cosmology stunning ingenuity seems to have coordinated chemistry and physics' natural laws, not least when it comes to bio-friendliness.**

No doubt, in any case, we're here because the universe is as it is. That's no surprise. How came it so? Was its initial condition chance or not? The universal body is sharply defined by a precise set of over thirty interdependent settings. *Their values combine to generate a universal pin-code that was either preordained or at least intrinsic in primordial projection. Indeed, the dials are set for the sun, earth, you and me to an accuracy computed by Oxford mathematician Roger Penrose at 10^{10} to power 123![47] If true, that cuts chance completely out. Erasure of coincidence.* The probability of your bullet hitting a nail-head at the other end of cosmos first shot is vastly greater. Mathematicians consider odds longer than 'only' 10^{50} against to be zero. That is to say, there is statistically no chance whatsoever that cosmos and its dependent life are accidental. **The Penrose computation, if valid, indicates that odds against the observed, law-abiding universe appearing by chance are stupendously astronomical; and the facts appear to support his calculation**. The consequence of such statistical annihilation kills off theories of origin by chance.

If chance is mathematically intolerable then whence? Where is the projector; what's the transcendent nature of projection's source? Natural Dialectic's 'holographic' edge is everywhere; it's 'super-posed' on physics, omnipresent but invisible because it's metaphysical. **It is the place where theme is turned to individual instance, principle is practised and where physic with its metaphysic meets.** Space,[48] time[49] and local things are *within* universal mind. **Mind's archetypes project our world; archetypal memories, potential matter, are the essence of our physic; they inform, unchangingly, material being. Immaterial information holds the world, physical and biological, together. It is by archetype conserved.**

This archetype (potential energy/ matter) can be thought of as letters in a universal alphabet, as bits, bytes and routines of a computer program, an alphabetic world. Particles and forces form, as physics well observes, the basic letters, punctuation and the grammar of our cosmo-logic's script; or even better (since vibrations/ wavelengths/ frequencies correlate with forces and particles), notes whose harmonics compose the cosmic opera. These principals can be conceived as carriers of archetypal code that compose the universal stage. The behaviour of non-conscious substance is thus metaphysically controlled.

Principles are concentrated information and, therefore, metaphysical power. They describe a source of order or a guiding force for local instances. As rules of a game are invisible on the pitch, nevertheless they substantiate its practice; *principle is universal, expressions of its practice individual*. **An archetype is a form of principle**; in this case metaphysical archetype governs the character of its physical expressions; it runs all players in the mighty, automatic game.

So, for the second saying, is it possible that archetypal forms are vibrations and thus, given the cohort of archetypal principles, cosmos is harmonically organised? *That creation can be thought of as an energetic vibratory scale in which its various bands are 'registers' or 'octaves'?* Such structure would allow that sound at a lower, slower, deeper level (say, physical) is played within the same overall composition as higher, lighter psychological forms. **The overall program, or musical score, could amount to universal opera.**

'Magna opera Domini exquisita in omnes voluntatis eius' is inscribed on the portal of the Cavendish Laboratories here in Cambridge. *'Great are the works of the Lord, carefully studied by all who so desire'* is not a maxim that has deterred a galaxy of scientific luminaries making profound discoveries.

Such opera is, of course, a reflection and a resonance of its transcendent order. In dialectical terms its Composer is Ideal and its Creator, at the Centre of Nature, First Cause. Song will, as every musician since Orpheus has known, pull you straight to the heart of things, to the centre of life. This is not a mathematical perspective but is it why Sir Brian Pippard, physicist and a Cavendish Professor at Cambridge University, wrote, "A physicist who rejects the testimony of saints and mystics is no better than a tone-deaf man deriding the power of music"?

(Raj) Active Energy

In *top-down* terms, we've seen, the vertically causative level of physical cosmos is metaphysical - archetype. In fact, another element that's metaphysical occurs - extent. This, Einstein might confirm, is compounded of two forms of formlessness - time and space.

In Natural Dialectic's spectral view, within this pair exterior appearance stems from subtle inner bands; creation issues, layered, from within. Now, therefore, moving out from metaphysic physics starts. We enter the kinetic phase of active energy, the world of quantum physics we call matter-in-principle.

Our world is built with no essential substance and, like music, out of structured energies. Yet this diaphany, claim physicists, is more substantial, subtle and more real than gross effects, a universe of bodies that we sensibly proclaim our truth!

However, from a *bottom-up* perspective, energy and its material agglutinations make up *everything*. Physic is the root from which illusions of the mind and soul have sprung. **Is energy eternal or has it any cause?**

Eternal matter? There was a theory of steady-state.[50] This has been rejected. Divine fallacy (god-of-the-gaps) is one thing but instead materialism recommends its own alternative, unholy artifice. It divines an unplumbed, endless gap called multiverse whence some titanic clash sparked our own universe; pure speculation is invoked to keep materialism on the road. Still, energy is something and, if not eternal, must in whatever form have had a start.

Big Bang
A Miraculous Projection

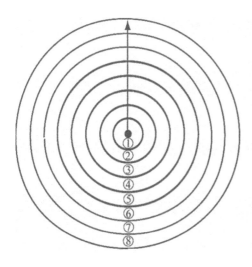

① naked singularity

② 'quantum fluctuation'; quantum pair production

③ inflation

④ mass-making 'drag' field; plasma; subatomic particles and nuclei

⑤ hydrogen/helium gas

⑥ embryonic galaxies

⑦ stars

⑧ solar systems; planets and moons

Therefore, have one. Current science lacks a reason for emergence into physicality. Its cosmogony tends towards the notion that for no reason, nowhere, absolutely nothing suddenly began 'exploding'. *In other words it relates the phenomenal start of a horizontal, age-related chain of cause and effects conceived of as a 'big bang'.*[51]

Big bang is, in this horizontal sense, a magnificent story of creation with a physics-only slant. Bang! Boom! Simplicity itself. **It might be, as we observe from 'inside' physicality, start's smoking gun.**

Therefore what headache, even to a physicist, could such a grand yet simple vision of creation present? Sadly, pass the paracetamol. Of nothing-physical, you ain't seen nothing yet.

Unseen but fecund void, called archetype, is where *top-down* causation metaphysically, that is, pre-physically begins. *As flower from seed so, in holistic view, exterior expression stems from subtle, inner code; creation is developed, stage-wise, from within.* **Cosmos not chaos sets initial conditions up. <u>Called potential matter, archetypal files are seen as hard a metaphysical reality as, say, particles are physical realities</u>.** Plan, principle, directive. Such potential comprises program(s) naturally stored in cosmic memory - simple in terms of inanimate physical 'law' (of particles and forces), complex in terms of animate structure/ function/ behaviour.

tam/ raj	*Sat*
physical matter	*Potential Matter*
action	*Physical Absence/ Void*
program played	*Archetypal Program*
lower syntactic level	*Upper Syntactic Level*
↓ *tam*	*raj* ↑
matter-in-practice	*matter-in-principle*
gross/ extrinsic constraint	*subtle/ intrinsic motions*
data item/ physical form	*data item/ physical force*
classical bulk	*quantum flux*

This stack and the one below dialectically express emergent levels. They help to show how archetypal 'nucleus' informs the way that energy precipitates, how certain energetic possibilities are locked into particularities of form and, by thermodynamic generality, material-expression-in-principle (the quantum substrate) and in practice (aggregates in three main states) are non-conscious phases of creation's shell.

	grade	*time*
sat	*pre-physical latency*	-
raj	*quantum micro-level*	*Planck/ quantum era of high energy and quick time*
tam	*classical macro-level*	*era of low energy/ bulk aggregates and slow time*

This stack's grades reflect the way 'big bang' is thought to have expanded, cooled and thus precipitated out the world. They may conform to a *top-down* projection of the 'vertical' and transcendental kind but equally can be construed as products of causation that is horizontal. That is to say, they conform to materialism's sense of progress 'light-to-heavy', 'simple-to-complex' and so forth. How, then, did particles and elements spill from a cosmic egg? Did chance really generate the cosmic forms or not?

Bottom-up, grant that photons, quarks, protons, electrons and the other fundamentals rose automatically. They rose from sets of rules spontaneously in place. It *must* have happened in a mindless way.

> *'Egg' is an apt metaphor for outworking from within; but development from an egg, far from being a random process, is preconditioned and involves precise codification.* ***And codification, demanding forethought, is a product of mind.***

In physic's realm the automated factory works witlessly but, as we've seen, instinct is a part of archetypal memory. By extension, the behaviours of cosmic body are defined by physical first cause, that is, a metaphysical precursor. Great network or machine are well-known metaphors for cosmic operation. Maybe the ancient and organic metaphor of cosmic egg and its programmed development has merit too.

And, if there is natural and vertical causation, there is mind behind. In Chapter 2 the illustration 'Order of an Act of Creation' showed how hierarchical creation works. ***Top-down*, from Natural Dialectic's spectral view, exterior appearance issues from interior.** Bulk matter follows from its inner, quantum nature. Whence does that nature take its form? Framing ideas accurately yet flexibly is syntax. Syntax, linguistic or mathematical, is convention or a legal framework in which symbols are ordered; its law naturally determines those structures allowed and those not. Thus the *upper syntactic level* (informant phase) acts as a filter through which order is communicated to and from the *lower (environmental, statistical or quantitative) level* (informed phase) of data items. In orderly emergence sub-atomic elements, forces and atoms of physics and chemistry can be construed as basic elements of code. Such code would dictate, through the agents that a study of phenomena elucidates, the way things naturally turn out. Quantum particles have been analogised to alphabetic notes/ marks/ letters of a cosmic language. ***Looked at this way cosmos is a Grand, Dynamic and Encoded Text.***

To complete the cycle we need to turn to creation's edge, the world's periphery.

This is the end. *Passive energy* is known as the classical phase of bulk matter. Such matter-in-practice is considered to be in locked, exhausted phase. Its 'bonds' include molecular formations, plasmas, gases, liquids, solids and, of course, all study related thereto. It represents projection's furthest radius from Source. The subjects of physics and chemistry deal exhaustively with this level of creation.

For this section let the final word rest with Max Planck. Planck not only first read, recognised and published (in the *German Annals of Physics*) the start of Einstein's revolution, the latter's Special Theory of Relativity. He also pioneered quantum theory, the second pillar of modern physics whose study is sub-microscopic, sub-atomic phenomena - the *matter-in-principle* of Natural Dialectic. Actually, his foresight may have ushered in the next revolution in human understanding which has already begun to focus on the primacy of the 'unscientific' informative co-principal as opposed to its energetic coordinate. He asserted that the discovery of truth can only be secured by a determined step into the realm of metaphysics and, at a lecture in Florence (1944) called '*Das Wesen der Materie*' (The Nature of Matter),[52] said:

"As a man who has devoted his whole life to the most clear-headed science, to the study of matter, I can tell you as the result of my research about the atoms, this much: *there is no matter as such.* All matter originates and exists only by virtue of a force which brings the particles of an atom to vibration and holds this most minute solar system of the atom together...*We must assume behind this force the existence of a conscious and intelligent Mind.* **This Mind is the matrix of all matter.**"

Indeed (The Observer 25-1-31 p. 17), "I regard consciousness as fundamental. I regard matter as a derivative of consciousness."

Perhaps Max Planck would have appreciated Natural Dialectic.

Chapter 5: Biology

Before the 1950s biology was a matter of 'outward' observation, copying, classification and, for about 30 years, some 'elementary' biochemistry (of vitamins, proteins and a suspicion, no more, that a substance called nucleic acid was a central component of all living things).

After 1953, however, things really took off. The structure, operation and 4-letter alphabet of a superb computer language was discovered. *DNA* is to biological form what computer chip is to an AI robot - except far superior in operation in that, like an entirely automated factory, it can cause the reproduction, growth, maintenance and repair of itself and the form it inhabits.[53] *DNA* codes, through an intermediate called *RNA*, for protein. Above are ranged discoverers Francis Crick, James Watson, Rosalind Franklin, Maurice Wilkins and 'runner-up', Linus Pauling.

A founding father of molecular biology was Malcolm Dixon. He was followed by Fred Sanger and Max Perutz who, again here in Cambridge, led the field in sequencing amino acids that compose a protein. In the same lab at about the same time Sydney Brenner realised that *DNA* was a language. Actually it is a digital quaternary (as opposed to binary) code; and, on top of that a double code. It may even be that, as well as epigenetic considerations, some sections of 'inside information' may read productively either way; such palindromic sense would require a very high level of ingenuity to formulate. However this may be, now *DNA* itself can be rapidly sequenced. Soon we shall have a library that includes the texts of all living organisms.

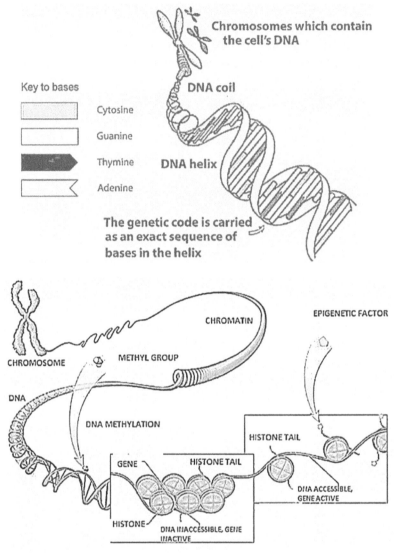

Of course, epigenetics - codes governing codes - amounts to extra informative dimensions.[54] Hierarchically arranged, its mechanisms build a towering, high-grade data structure. *A responsive genome incorporates multiple, overlapping and interactive code in a mode the IT profession calls data compression.* **How can first-class data compression arise without a prior informant?** Ask any software engineer. *Why should DNA and its controllers, packing information just as tight as any hard drive, be different?* **In fact, its chemicals express**

each bio-program. _DNA is nature's chip._ So the basic bio-issue is, unrelentingly, the source of code.

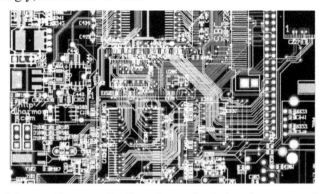

The chip specifies protein. Protein synthesis involves transcription and translation into protein using molecular industrial units, some composed of many subunits. Transcription is a multi-step process employing polymerase and other molecular machines. Translation, at the other end, is also multi-stage; many machines, including multiplex ribosomes, cooperate. **No theory has been proposed for the evolution of the critical, immediate and highly serendipitous linkage of protein with nucleic acid through the medium of such a suite of requisite, complexly-formed and yet precise t-RNA translators and corresponding synthetase tools.** Without this series intact nothing could start. So it goes on. _At least 100 different proteins, each coded for and synthesised by the very machine of which they are components, are used to convert DNA instruction into protein product-hardware._ Of the whole system the late Professor Malcolm Dixon, a founding father of enzymology (the study of enzymes) at Cambridge University, wrote, *"The ribosomal system under which we include DNA and the necessary co-factors, provides a mechanism ... for its own reproduction but not for its initial formation."* Now, as well as source of code we have to ask how, from the first cell, its programs incomparably output the goods!

On these matters naturalism is tenaciously defended. The **Primary Corollary of Materialism (see Chapter 1)** states that life forms are the product of the chemical evolution of a first cell (see Glossary); and following that, the neo-Darwinian theory of evolution. Evolution is, in this sense of the word, change in the heritable characteristics of biological populations over successive generations. Changes are caused by the main mechanisms of sexual recombination (after sex evolved) and mutation acted on by a filter called natural selection. The creation of life on earth is thus an absolutely mindless, purposeless process.

Biologist Theodosius Dobhzhansky is famous for coining a popular mantra: 'nothing makes sense in biology except in the light of evolution'. **But the actual, iconoclastic fact is: 'nothing makes sense in biology** except in the light of information.' You may drag evolution in on information's tail but it is simply a fashionable word used in biology to mean several different things. *In reality, codes and signals run the bio-show.*

In other words, not evolution but information is the basis of biology - code that yields irreducibly complex structures and cooperations. So Darwin, seeking an origin of species and ignorant of biological *sine qua non*, asked the wrong question. *It is the natural origin of information needs be rightly explained.* Every cell in every body depends on the successful integration, by program, of complex factors without which it would not work. This is called its irreducible complexity and is tightly linked to minimal functionality (meaning a mechanism must fulfil its promise to the extent that an efficient, acceptable level of performance is achieved). Physical elements and natural forces could not make a cup of tea in a billion years, let alone its drinker.

The issue is not one of religion or opinion but science and logic. Information is, recall from Lecture 2, demonstrably an immaterial factor. It is metaphysical. **Therefore, the basis of biology is metaphysical.** And such holistic perspective (including both material *and* immaterial elements for consideration) renders the notion of chemical evolution logically illogical. *The promissory faith in obtaining a 'first cell' by chance (as opposed to 'chance' guided by great scientific ingenuity both in mind and in the lab) is irrational.* **E cellula omnis cellula**. Only from a parent cell does daughter come. **No exception has ever been found to this rule so that it is called The Law of Biogenesis.**[55]

Can a book robotically assemble its own plot? Biological code accurately informs the manufacture, maintenance and reproduction of specific, complex cells and bodies. Ask any IT specialist or inventor if non-conscious, aimless nature could better produce his programs or his mechanisms. Time is not the issue. Mind is. To try to obtain a program by chance is, in fact, a most irrational venture. <u>**For this reason, since information is the basis of biology, the theory of evolution is, in principle, fundamentally and demonstrably absurd.**</u>

In order to sensibly discuss an inclusive, holistic alternative, let's suggest a brief but broadened theory of biology. We've already looked at archetypal memory, first cause of all things physical. How, amid the specific, functional and fully codified complexity that is organic form, is this expressed?

In the green diagram we follow the triplex, dialectical order of expression. *Information precedes.*[56] It is prior and anticipates. Information is the potential and *sine qua non* for the action that consequently issues orderly. Indeed, it was suggested in Chapter 3 that every cell involves sub-conscious mind in the form of its typical mnemone, that is, its archetype.

Information comes, as previously explained in Chapter 2, in two forms - active/ conscious and passive/ unconscious. It involves both

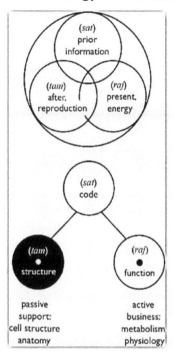

A Dialectical Plan of the Way Life Works

tam/ raj	Sat
gross/ subtle expressions	Potential Phase
subsequent order	Archetype
codified outworking	Message
subsequent effects	Prior Information
↓ tam	raj ↑
informed structure	informant function
specified build	signals
organ	metabolism
individual part	systemic communication
external/ gross	internal/ subtle
end-products	action

psychological and biological components. **The unconscious, informed case** shows as archetype and soft, bio-logical machinery. The pair includes archetypal instinct and morphogene; reflex balance, called homeostasis, by way of nervous, hormonal and other systems; cybernetic metabolism controlled by preordination in the form of genetic code carried chemically by *DNA*; and muscular organs of action and response. We discussed archetype in the sections on psychology and physics. In regard to the latter, the assertion that particles and forces constitute an alphabet, punctuation and grammar of a fine-tuned text is sometimes met with a shrug - things simply happened as they are. *However, considering biology's complex, codified operations you have to shrug this shrug off; the view changes dramatically.* If cosmos is fine-tuned why not, with very intricate and purposive complexity, biology as well? Suffice, at this point, to reiterate that if I asked you for *physical* proof of mind in the chair you are using, you would rightly dismiss the question. Yet even a simple chair has the mind of its designer definitely there. Similarly, mankind may eventually come to understand every last quantum of soft, codified bio-machines but will the chemistry be all? The most important element, the source of information needed to construct them in the first place will have been ignored. Such an explanation is, as Polyani noted, in no way complete.

Life's process is one of dynamic equilibrium. Equilibration. Its goal is cybernetic balance in accord with pre-set norms. Metabolism, being totally information-dependent, works with reference to precise, incoming messages and equally precise genetic response. Such fixed response is indexed, switched and flexibly monitored by non-protein-coding and epigenetic factors; and also by interconnected nervous and hormonal systems. On the conscious, psychological hand, a flux of desires creates a moving set of targets whose equilibration (or neutralisation) is reached in satisfaction.

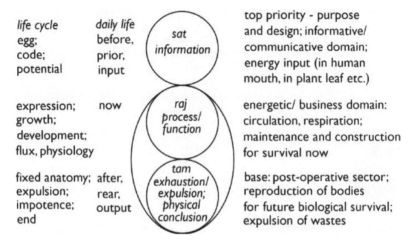

life cycle egg; code; potential	daily life before, prior, input	*sat information*	top priority - purpose and design; informative/ communicative domain; energy input (in human mouth, in plant leaf etc.)
expression; growth; development; flux, physiology	now	*raj process/ function*	energetic/ business domain: circulation, respiration; maintenance and construction for survival now
fixed anatomy; expulsion; impotence; end	after, rear, output	*tam exhaustion/ expulsion; physical conclusion*	base: post-operative sector; reproduction of bodies for future biological survival; expulsion of wastes

Life is, seen this way, an incarnate flux of order due to information. **In short, organisms are information incarnate**.

Bio-potential, information, is expressed by specific action. **Energetic actions** provide for survival now. They involve, metabolically, photosynthesis and respiration which in turn promote cell biochemistry, trans-membrane voltages, physiological processes and, on the large-scale, (nervous) sensation and (muscular) motion. The character of all function is energetic.

Biology in Brief: Information Plugged into Structured Energy

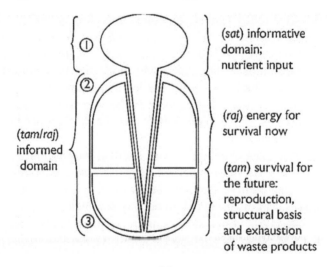

(*sat*) informative domain; nutrient input

(*tam/raj*) informed domain

(*raj*) energy for survival now

(*tam*) survival for the future: reproduction, structural basis and exhaustion of waste products

Structure, whose character is solidity, represents the outermost, fixed (or flexibly fixed) realisation of shape. Its base domain is energy's container, a fixed container for the expression of internal, orderly flux. In other words, a 'phenotype' is a peripheral aspect whose body both reflects and fixes the shape of inward information and energy. The end-product of structural development, maturity, is reproductive. The cycle starts again.

Biology in Brief: *Top-down* Man

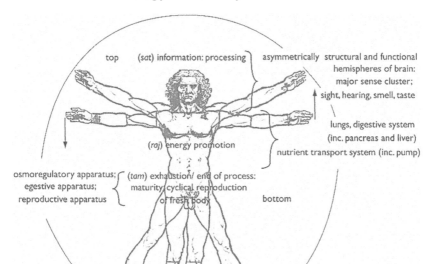

For Natural Dialectic, therefore a cell or multicellular body is primarily information, only secondarily biochemical. And, as we saw in Chapter 3, the structure of your own body is an expression reflecting the cosmic fundamentals.

We have nuclear super-computing but, in the energy department, the central executive is homeostasis.[57] It is based upon the geometry of periodic action, a circle and its extension, vibration. Vibration round a norm is at the root of biological equilibration, its dynamic equilibrium, its energetic stability or superb balancing act. The controls are set for balance; these controls are always, in principle, triplex. They involve mechanisms of sensor, processor/ controller and effector. *All homeostasis, in any cell, must involve these three deliberate components; and each must be codified before it can be built and work.*

Why, in the face of physical entropy, should equilibration be the goal? Why, in spite of nature's fall towards exhaustion, breakdown, garbled chance events and death, should a tight-rope for survival be so evidentially codified? Indeed, why should a group of atoms ever 'want' to survive? Is not the whole cybernetic business of homeostasis for stability of operation? **Cybernetics intrinsically involves anticipation, purpose and a goal.** Biological energy is, through documented process and specific agents, strictly guided; it is, in a word, completely informed.

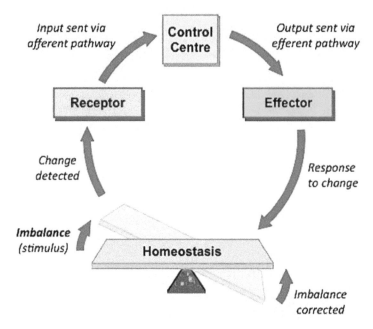

Code anticipates. Signals communicate. Information operates with feedbacks which ensure dynamic systems 'stay in line'. Therefore, to find the causes of 'downstream effects' you have to travel to their 'upstream' cause. *Such cause is not natural selection. Nor is it mutant lesion of a gene or chromosome.*

Hierarchical information-structures are conceptual. In other words, visible characteristics called phenotype depend on molecular action that, in turn, depends on genotypic and other pre-programmed information. *Life is so obviously purposeful in character that genetic chemicals are ascribed all kinds of animistic properties.* They 'compete', 'organise', 'express', 'program', 'adapt', 'select', 'create form', 'engage in evolutionary arms races' and even, lyrically, 'aspire to immortality'.

The origin of hierarchical order, cyclical flows of information and integrated function is always purpose. The origin of purpose and its attribute, meaning, is mind. *We might, therefore, reasonably infer that the basis of meaningfully informative, functional and structural biological hierarchies is mind.* So biology's first, prior set of hierarchies is identified as informative.[58]

After informative hierarchy (code to protein and chemicals, thence to functions and structures), its second set of hierarchies is energetic.[59] Just as a brief example of informed complexity of mechanism note four illustrations.

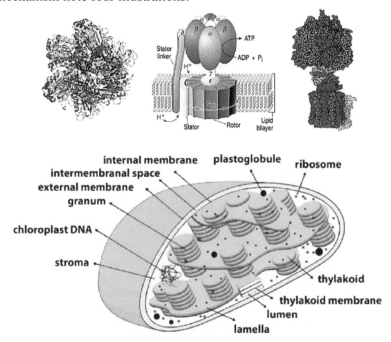

The top three pictures show one component of respiration, a 'dynamo' called ATPase. It is used to 'burn' the sugars created in an organelle found in plants and phytoplankton called a chloroplast (fourth illustration). Energy metabolism is composed of a duo - photosynthetic build-up and respiratory breakdown of specific molecules. We now know that its complex engineering includes, at the initial stage of photosynthesis, quantum biology as regards the resonant energy transfer of electron excitation in a coherent, most efficient way.

Replication, energy metabolism and reproduction. Whence came the information for all these essential systems? *Bottom-up*, the problem's to convince yourself that, granted a vastly complex starting-point, the computation of cell chemistry chemistry haphazardly 'improved itself'. What mindful systems flow-chart could you build to demonstrate this mindless possibility?

Top-down, programmers know that, from a main routine, switches branch to sub-routines and, when a sub-routine is done, the process cycles back to start again. **These routines are modules**. This conceptual character of algorithm, this repetitious use of switches and blocks of modular code is just what coded bio-systems show. It is how nature's life-forms, full of reason, always work.

One would, therefore, predict research will more and more reveal signs of bio-logic to the point that, in 'live' computing, such complex permutations, integrated combinations and hierarchical sets of regulation will further squeeze then nullify the notion that celled systems ever cropped up accidentally, that is, evolved.

A computer is a mind machine.[60] *On this basis it is established that the cybernetic operations of a cell, an object as thoroughly material as a computer, superbly meet the criteria that pass it as a mind machine.* A cell is a mind machine. Its instruction manual is a program written up as 'genome'; and its systems hold semantic meaning as modules or entities of bio-form like, for example, you.

Various tailored sub-routines are called from a Main Routine. Suites of such modules combine as permutations around which different bodies are, under the coordination of such an Archetypal Master Routine, expressed.

Molecular biology's great strides are also heading straight towards the notion of an archetype. A similarity of genetic instructions is found to extend across a great variety of forms. For example, a high-level developmental complex of modules (called 'homeotic genes') may code for outline body-plan in very different organisms such as human, duck or fly. Or it may call subroutines for the construction of completely different kinds of eye - a normal gene from a mouse can replace its mutant counterpart in the fly; now an eye, a fly's eye not a mouse's, is produced. In other words, such genes act as 'go-to' switches in an archetypal program of development.

One could add much more but we now turn to a brief critique of the biological theory of evolution in terms of three tenets and five ingredients: speciation, natural selection, mutation, sex and development.

Perspectives on Three Central Tenets of neo-Darwinism - a Tabulation

		Bottom-up	Top-down
✓	true		
✗	false		
①	Abiogenesis	✓	✗
②	Variation (microevolution) by mutation and natural selection	✓	✓
③	Transformism (macroevolution) by mutation, natural selection or any other means	✓	✗

A half-truth is the most difficult to unravel. Elements of Darwin's theory are, of course, agreed by everyone. Everyone agrees that variation-on-theme continually occurs. Such variation is the result of sexual reproduction (which contains its own in-built lottery called meiosis) and, deleteriously, by genetic mutation. ***Top-down*, it can also the product of adaptive potential.**[61]

Remember what was said at the start of this Chapter? **The basis of biology is information. Therefore, Darwin asked the wrong question.** *With what we now know the question does not concern an origin of species but, fundamentally, the origin of information.* Nevertheless, let's take a look at speciation.[62]

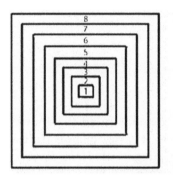

Vehicle	Body (vehicle of life)
1. carriage inside or outside vehicle	domain prokaryote/ eukaryote
2. type of propulsion	kingdom e.g plant, animal
3. motion in air, on land or in water	phylum subphylum e.g vertebrate
4. land: motor car, lorry, van etc.	class: mammal
5. car	order: primate
6. make: Porsche	family: *Hominidae*
7. Carrera	genus: *Homo*
8. species of Carrera e.g Cabriolet	species: *Homo sapiens*

Carl Linnaeus, the founder of modern taxonomy, thought a 'species' was a group whose members could interbreed. Darwin, as opposed to Linnaeus, preferred to 'plasticise' his definition of species to what he saw as its cause - gradual change. He wrote *"I look at the term species as one arbitrarily given, for the sake of convenience, to a set of individuals closely resembling each other, and that it does not essentially differ from the term variety."*

However, Darwin not only regarded varieties as incipient species but also proceeded to *extrapolate* on the hypothetical principle of unlimited plasticity. He *did* consider this, a process called micro-evolution with which everyone agrees, a minor stage of macro-evolution. In other words, he **guessed** that variation was a progressive rather than a constrained, cyclical process. The question is plasticity, that is, the real extent of variation.[63]

Such extrapolation, variation-*without*-theme, formed the very basis of his theory. Through materialism's prism life is, naturally, progressive from a simple start. This, unlimited plasticity, is unquestionably the current scientific mind-set. Yet, we'll see, even the part of 'simple' start (called abiogenesis or chemical evolution) is fraught with intractable problems.[64]

Bottom-up, Unlimited Plasticity

Top-down, Limited Plasticity

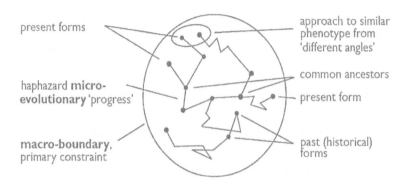

In practice, when genetics decisively demonstrates the limits past which a mutant does not survive it also demonstrates conclusively, through myriad experiments (e.g. with *HIV*, *E. coli* bacteria, fruit flies, malarial parasites, flowers and mice), that the macro-evolutionary principle of unlimited plasticity (or unbridled extrapolation) is incorrect. Ask a dog-breeder if he has bred other than a dog. Ask him whether anyone has ever, by intelligent selection, pushed canine elasticity into the production of other than crippled or still-born, monstrously deformed dogs. No-one has made another type of organism from a wolf. **This demonstrates a limited plasticity.**

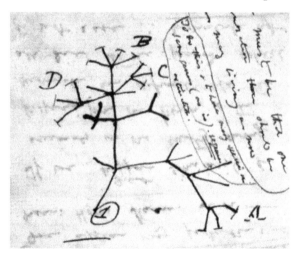

Darwin's inspirational doodle isn't right. The central assumption of

82

his tree of life is that homologous molecular or morphological patterns will, by comparison in different organisms, illustrate degrees of phylogenetic relationship. Hasn't this idea been uprooted, sawn and cast as dust by an 'onslaught of negative evidence'?[65]

Thus, where Darwin staked his claim on continuity, *discontinuity* (the primary prediction of an archetypal model) is what we actually find. Nobody is more aware than a *palaeontologist* that Darwin's 'interminable varieties' are missing from life's petrified family album.[66] And if, of course, the truth is that types are ring-fenced and discontinuous. Family trees are separate. Life's in fact a forest not a single tree.

Chemical evolution is demonstrably a non-event yet study attracts much taxpayers money (in the form of grants and university tenureships) around the world. Why? Because it is the seed, the root of evolution's tree. If the axe were laid here why not elsewhere? **For example, a couple of factors (in addition to *DNA* chemicals, replication and energy metabolism) had to be present correctly codified in the first cell - transcription and translation through combinations of enzymes and micro-machines such as ribosomes. Such process, multi-component, multi-stage and self-coded-for, must be working from the very start of bio-informative operations.** In this case, why is the rational notion of an archetype so unintelligent? It is only materialism and, especially, its subset of scientific atheism that need to damn it so.

The human genome could be stored on perhaps 1.5 gigabytes of space. Its 250,000 or so pages of close-written lines of letters would compose a book weighing 450 kilos. This code needs be accurate. At the other end of biological scale biologist Craig Venter defined a minimal genome by removing genes, one by one, from the already very small genome of a *Mycoplasma* bacterium. If the organism survived he discarded the gene until he was left with a minimal genome of 473 vital genes. But symbiotic *Carsonella rudii* has certainly lost the bare necessities of life since its present genome of 182 genes (~160000 'letters') is insufficient to replicate or transcribe and synthesize protein. A viable analogy for free-living *LUCA* might be ten times that amount, that is, 1.6 million letters (the size of a 600-page book) in the precise order of a program that will create specific molecules and shapes. Are natural selection and mutation up to the task of sufficiently 'improving' the survival chances of a highly speculative first cell?

First, historically, let's take Darwin's genetic editor called natural selection.[67] Edward Blyth (1810-73), a keen naturalist and 'Father of Indian Ornithology', published essays that first appeared in The Magazine of Natural History in 1835, 1836 and 1837. *These, which were read by Charles Darwin who afterwards corresponded with him, introduced the ideas of a struggle for existence, variation, natural selection and sexual selection.* These four are central Darwinian tenets and we might ask why, since in large part Darwin's work was based on Blyth's ideas, the former hardly acknowledged the latter's 'intellectual copyright'. And Alfred Wallace (1823-1913), co-founder of the theory of evolution and whose paper 'On the Tendency of Varieties to Depart Indefinitely from the Original Type' preceded (1858) and precipitated publication of Darwin's 'Origin of Species' (1859), noted that natural selection not so much selects special variations as exterminates the most unfavourable. Such extermination is, in fact, a stabiliser; 'fit' traits blossom, 'unfit' wither as an organism's own ecology (its niche) dictates.

So let's treat editorial selection for what it actually is and, at the start of our inspection of its 'mechanism', post three *caveats.*

The first, Blyth's, is that natural selection is solely a process of elimination. It involves the disappearance of 'unfit' organisms. **In short, deletion is not creation. Natural selection originates absolutely nothing at all.** It is, simply, a fateful finger hovering above a genetic delete button, no more than a name given to the lucky survival or unfortunate death of an organism or group of organisms in a particular environment.

Secondly, the effect of natural selection is, in the case of speciation, to have *reduced* the genetic potential of an original gene pool. Alleles are deleted or a pool split into separate populations. Information is not gained but lost. *For evolution, which needs not information ($\downarrow$) loss but ($\uparrow$) gain, this is entirely the wrong direction.*

Thirdly, it is false to conceive of natural selection as a kind of 'ratchet' that holds on to 'an extrapolation of improvements' at any level from nucleotide through protein to whole body shape. This is because without a prior plan to work towards there is no way for intelligence, let alone total lack of intelligence, to 'discriminate' a 'good' apart from 'bad' move as regards some novelty. *Indeed, it will breed out unwanted, nascent or non-functional characteristics.* **Only once an organ or a functional system (such as a beak or eye and associated factors) is complete and fitly working can natural selection act on trivial, accidental variation to that system. In a phrase, the 'mechanism' explains survival not arrival of the fittest.** It weeds the weaker but cannot create the fitter; it's as creative as a kitchen sieve.

In short, this mindless editor confers not change but stability by maintaining wild-type pedigree. Far from being a grand 'law of nature' natural selection is a trivial observation - an organism born defective does not reproduce. Its truism states the obvious. 'Weaker die, stronger live'. 'Who survives, survives'.

What about the mechanism that provides the forms that natural selection hones?

Mutation?[68] In 1865 Gregor Mendel presented his Theory of Discrete Units of Heredity. This was picked up by the Royal and Linnaean Societies in London and (since 'adamantine particles' are inflexible and anti-evolutionary) lost until, in 1901 de Vries discovered a way out - **mutation**. Henceforth random mutation became the innovator, the creator on which natural selection might, by killing, work. **Materialistic faith is vested in a core of unpredictability, a central lack of reason - mindless chance.** A scientific G.O.D, no less! A Generator of Diversity! By philosophical necessity current science takes its chance on chance. Order came about by chance. No telling how exactly, just vague imprecation. The story's scientific *aide-de-camp* is probability. Nothing is, perchance, impossible; the sole impossibility is that such a story is impossible. Materialism's Lady Luck, however weak, is elevated to almighty creativity, creator of life forms.

Of course, chance does impact life. In a few cases and contexts genetic mutation may help a species survive. Lucky chaps! *But can the Lady's empty womb bear any organism in the first place?* If mutation, a by-product of genetic operation over time, simply breaks or loses information from a gene then it *devolves* an old form not evolves a new. **At best it generates predominantly damaged variation on a predetermined theme**.

Nevertheless, mightn't mutations, if you found millions of 'beneficial' ones, transform one type of body to another in the way accretive evolution needs? *'Beneficial mutations' (BMs) are to genetics as 'missing links' to evolutionary palaeontology - critical.* **Not just The Primary Corollary but The Primary Axiom of Materialism and its whole panoply of philosophical, political and sociological speculation, not to mention the paradigm of modern science, hangs on their slender thread.** *The whole of secular academy depends, for its verbose existence, on this evanescent gleam of hope, a key to unlock all of life's diversity that's called a BM ('positive' or 'beneficial mutation'). Can BMs rise to such occasion? Everything depends on this.*

In fact, any *BM* would be invisible at the level of a whole organism and its small 'advance' overwhelmed by neutral or deleterious mutations long before the successive and exact chain of cooperative *BM*s needed for any novel biological system could 'evolve'.

Not knowing where you're going is a fundamental problem too. It applies to 'junk', 'neutral' or any other kind of *DNA*. What should a first or following *BM* be? One can only be defined within a preconception. What is 'good' or what is 'bad' is only so when valued in anticipation of that end. How can you even define a *BM* if you lack direction? The atoms of a system do not understand it. Mindless evolution's natural selection doesn't know. And if there's no such thing as target how can 'progress' gradually mint a system with its integrated working parts? **With such lack of logic bang goes the irrational *PCM* and therefore bang goes *PAM* as well!**

The fact is that *DNA* represents, effectively, an organism's program. It has, however, never been shown that a coding system and semantic information could originate by itself through matter. IT's information theorems predict this will never be possible. Yet the basis of biological program is code. Code is always the result of a mental process. If code is found in any system, you might conclude that the system originated from concept, not from chance - especially if that code is optimised according to such criteria as ease and accuracy of transmission, maximum storage density and efficiency of carriage (such as electrical, chemical, magnetic, olfactory, on paper, on tape, broadcast, *DNA* etc.); and if, above all, it works and orderly instructions are unerringly responded to. Code and incoherent chance are chalk and cheese.

Randomness of any kind is reason-in-reverse. Whatever is encoded is intentional. *A coder takes no chance. Randomness is eliminated.* By definition, mistake or randomness degrades information; and the job of any editor (programmed *DNA* editors included) is to eliminate interference, 'noise' or mistake. **Chance neither creates nor transmits information. On the contrary, accidents always (unless accidentally reversing a previous degradation) degrade meaning and, by degree, render information unintelligible**.

Does *DNA* really operate like a digital computer program?[69] There are close resemblances. And every systems engineer knows that a change to his construction (whether accidental or deliberate) either brings things to a grinding halt or causes ripple effects that, unless properly balanced and calibrated with cooperative parts elsewhere, degrade the functional intention of his design. Similarly with software. Random changes without reason or systematic realignments have, at best, no effect or, at worst, bring catastrophic failure. They never improve it! This is exactly what we find in the case of biological mutations. They are not a good choice of creator to underpin the theory of evolution and its materialistic creeds. Indeed, they intrinsically pronounce it incorrect.

Yet, from ignorance, it was touted throughout the 20th century that we knew such bio-programs whose operation we still hardly understand evolved by chance mutations acted on by natural selection. Chance mutations '*must have*' caused the code with its adaptive, switching systems but, after this appeal, it is never explained exactly how (in what combinations or algorithmic steps) or why such mutation was, although at that stage neutral or harmful (i.e. useless or worse), on the way towards some 'improvement'. **This is not science. It amounts to the multiple invention and repetition of just-so stories.**

Yes, variation-in-action certainly exists. All agree upon the limited plasticity of micro-evolution. And on adaptation to fresh circumstance. Could not, however, an **adaptive potential** (Glossary) be written into genome and its epigenome so that it would not be mutation but a flexibility of program that produced coloration changes, different beak sizes and so on? Various mechanisms have been suggested, none of which , however, come near to explaining the *innovation* of any integrated, functional programs, organs, systems or, beyond variation-on-theme, transformist macro-evolution.

In science you experiment. Can you test evolution in the past? In his book 'The Edge of Evolution' Michael Behe demonstrates that nature has empirically tested Darwin. Do you want numbers showing how, by gradual mutations, life might step-by-step evolve its family tree? *HIV, E. coli* and malarial parasites satisfy the numbers game. Virus, bacterium and eukaryotic cell have reproduced, mutated and should have evolved through sufficient generations with sufficiently large populations to indicate whether neo-Darwinism's engine, random mutation, can bear the weight of a theory that would have it gradually innovate parts and body-plans by gradual but cumulative, useful steps - or, buckling, is crushed by numbers.[70]

In short, we understand that genetic information is at the root of biological form creation and maintenance but not how the program 'knows' how to build objects and events that make precisely the right parts at the right times in the right quantities and places. We do not know exactly how this automated factory program generates development (which involves anticipation) of any tissue, organ, system or whole body. When we do we will be able to say that we thoroughly understand this or that machine in all its aspects. But will this mean, as a materialist might claim, that we completely understand?

As Michael Polyani argued, machines are irreducible to physics and chemistry. *They are irreducible because they involve immaterial purpose, the stepwise development of a plan of implementation, a directed cohesion of working parts and, of course, the thoroughly non-material anticipation of an operational outcome.* For example, to completely analyse a bicycle does not mean you have completely understood that form. You need to include its purpose, conception and technological development, that is, the vertical causation of mind in it. Simple physico-chemical analysis would never by itself obtain more than a fraction of the whole truth.

Seen in this light, micro-evolution is a prejudicial word for variation-on-theme, biased because it implies the existence of an extended process for which no hard evidence exists. **The reverse. Macro-evolution is a theoretical phantom**. **In this view, evolution-in-action is simply variation-on-predesignated-theme**; this variation is always constrained by working systems already in place; and it is either coherent, due to in-built genetic potential or incoherent by neo-Darwinian mutation. *Variation proliferates; the special theory (STE) is right. But the general theory of grand macro-evolution (GTE) is not.*

Energetic matter knows no future. It cannot program, plan or anticipate. Yet, informant egg to informed adult, reproduction shows anticipation. Target. Purpose. Reproduction[71] defies entropy and death. It means the type survives. Let's take a look at the third of biology's hierarchies - exhaustive. There is exhaustion of waste materials but also, by reproduction involving the lower quarters, expulsion of fresh bio-forms.

There are three main types, with many detailed variations, of the way that organism's reproduce. Due to lack of time we take, of these, meiotic sex.[72] The biological point of sex is variation-on-theme. It is neither to add nor subtract but to shuffle an organism's cards into new permutations and thereby deal new hands in an old game. In this case two become a different one. Sex involves clear anticipation, targets and complex programming - not features normally attributed to mindless matter or to chance. How did it arise?

If you wanted, theoretically, to design complementary sexes you might opt to specify each module using thousands of genes; *alternatively, and far smarter*, *you might conceive a main routine which included a 'gender switch'*. *Each type of life form that exhibits sex is conceived of as a neutral whole divided into polar male and female sexual halves*.

The Archetypal Polarity of Sex 1

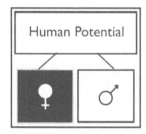

The sexual archetype (<u>not</u> a common ancestor) and its *DNA* coding are fundamentally hermaphrodite and incorporate both male and female potential. Such hermaphroditic concept's switch would trip a male or female line; it would call gender-informed sub-routines. In principle, therefore, at a mere flick the balance would be tipped. A hierarchical cascade of emphasis would, either way, ensure dimorphic forms occur. **This skilful concept, this consummately programmed switch is precisely what we find in bio-practice.** From the same cells grow, in each gender's case, male and female parts. For example, sensibly perceived, male breastless and female breasted nipple are, as penis and clitoris, examples of the differential expression of the human archetype. Indeed, dimorphic *and* hermaphroditic (but never multimorphic) sexual algorithms are found in plants and animals. From hermaphroditic potential derive uni-morphic hermaphrodite, di-morphic male/ female and other forms of expression such as, by successively alternating sexual *and* asexual forms, even manage to polarize the reproductive archetype itself.

The Archetypal Polarity of Sex 2

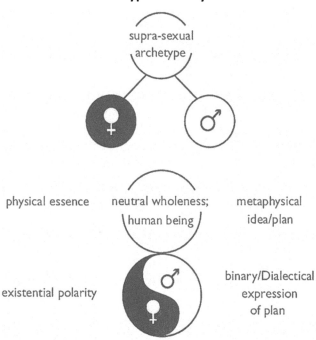

You know for certain, therefore, the same program can by switch deliver, for example, sperm (including the set of over 600 proteins that compose its eukaryotic flagellum) and egg (with her own specialities). Of course, genetic switching *aberrations* (e.g. intersex, abnormal hermaphrodites or trans-sexual tendencies) may occur as well but, normally, sex is implicit in a neutral archetype's potentiality.

And as for eggs, did an egg precede its adult?[73] You can't see a memory but have you ever seen a physical idea? Ideas are potential; from potential outcomes are evolved; an egg, packed with symbolic information, is as near to 'natural idea' as you will get.

Did an adult precede its egg? Did fruit precede its branch? We noted that reproduction looks to the future; and thus sex *anticipates*. It is a conceptual process engaging machinery of irreducible complexity to achieve a target - generation after generation. Survival of the kind.

Egg and Adult Together. What was noted in Chapter 2 bears repetition. **Wherever an apparent 'chicken-and-egg' situation crops**

up the puzzle is resolved by the introduction of purpose, design, information and mind rolled into one - teleology.

As far as chicken and egg are concerned, therefore, the simple answer of an information technologist is that neither came first. One did not precede the other. You make them together with the same object in mind - in this case a self-reproducible biological information system or, in other words, a living being. **You are one half of a metaphysical idea called 'human being'.**

***Meiosis**, which underwrites the sexual reproduction of some single-celled and nearly all multicellular organisms, bears the hallmarks of a choreographed routine designed to extract maximum variation-on-theme at minimum cost in labour and materials.*

Meiotically generated gametes then, by means of and within a highly specified and complex context, fuse; and their offspring, a single-celled zygote, unfolds according to complex but codified, specific algorithms. A hierarchical developmental archetype eventually realises its goal, its reason - re-creation of the next adult generation.

The Order of Development:
Developmental Stereocomputation

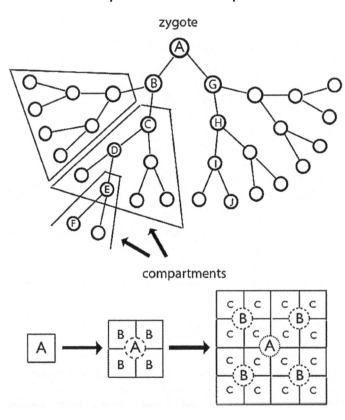

90

At this point we can ask how an engineer might design a self-reproducing machine and ask if nature has preceded him in his intelligent logic. How might the least demand be made on the tissue or strength of a parent while at the same time encapsulating its potential? <u>A brilliant, optimally economical idea would be to reduce the parent to a single cell and then, from the symbolism of this cell, build up a new adult.</u> This is exactly what happens in nature. Each individual adult is 're-potentiated'. Its body is reduced almost entirely to a symbol, a directory, a coded book of what might be. Its spring is re-compressed into the top-level potential of an egg or sperm.

Hierarchical Control

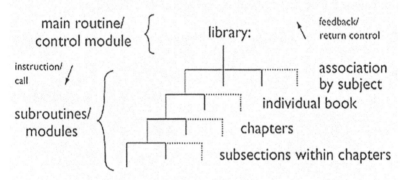

top-down or modular programming:
hierarchically nested command and control

main routine/ control module — library: — feedback/ return control

instruction/ call — association by subject

subroutines/ modules — individual book

chapters

subsections within chapters

hierarchical information structures are always conceptual;
this simple example applies to the hierarchical classification
of a library but could equally apply to the structure of a
complex computer program or, from outline to detail,
the integrated construction of a machine.

Developmental biologists uncover switch-arrays and associated, complex, hierarchical cascades of regulation.[74] They reveal top-level 'master genes' or 'tool-kit factors' that determine which parts of a body will develop out of others. These blocks of levers clustered as a signal box that rules developmental lines remind you of a railway running from a start to terminus. With main and branching sub-routines they remind you also of computer software structures. Regulatory logic-clusters are found, it will predictably transpire, in all multicellular animals. For example, in four distinct blocks in vertebrates (called *Hox* gene clusters) each gene is responsible for triggering a cascade of sub-routines that will supply the right materials in the right place at the right time to construct a given segment of the animal in question. Obviously, therefore, a *Hox* gene for a worm will define (or, in computing terms, call) different or deeply differentiated organs (sub-routines) to ones for a fly, a horse or a human. In other words, the same genes can be responsible for initiating the development, in terms of systems or organs, of different outcomes. In one instance a gene may specify for a tail, a coccyx or the rear part of a fly, frog or grasshopper; and the gene that triggers development of your eye would, if transferred to a fly that lacked it, cause its blindness to be eyed - with a fly eye not a human one, of course!

In short, a dialectical bio-hierarchy is not horizontally caused by genetic accidents through time, that is, by a successive 'origin of species'. It is, instead, the product of vertically caused genetic program (with its preconceived metabolic and physiological subsequences) designed to develop and maintain organic structures. If there is objection to this idea it is not scientific but due to the philosophical prejudice of naturalism and its consequential, mindless author - chance.

For example, how does chance evolve a path that targets goal? How, entirely ignorant of end-game, did as-yet-useless intermediates survive? Missing links die of incompetence thus how did any metabolic pathway prophesy its own construction or feedback control; how (as, analogously, in the logical, consequential proof of a mathematical theorem) did thousands of correct steps on the path to reproductive adulthood arise by accident? Evolutionary 'must-have-somehow', 'just-so' stunts are pulled continually except, with metamorphosis, the bluff is called spectacularly. Butterflies have always thrown Darwinism, in an arm-lock, in a flap and on its back.

Back-to-front. The transformation from egg to beetle, bee, fly or, more strikingly, a butterfly illustrates a pattern of development that defies a gradual, practice-to-principle evolutionary explanation.

Entirely different-looking phases, each perfectly formed for function, serially erupt. First *egg*; then, for a 'childish' *caterpillar* to become an adult moth or butterfly, it eats and moults. It keeps moulting exoskeletons (which are flexible like cat-suits and yet give it shape) for larger ones that form folded underneath the smaller outer sheath. At the last and largest size skin is shed by delicate manoeuvres revealing a cocoon, a *chrysalis* in which the future hangs. Then caterpillar body parts dissolve and build again into a butterfly called the *imago* that, emerging after several days, inflates its lovely wings by pumping blood into their veins.

Could humbly waiting as a caterpillar eventually evolve a program (saved as what are called 'imaginal discs') for pupal development? In what nascent cocoon could enzymes 'know' how far they should dissolve a caterpillar's parts before re-building to transfiguration called a healthy butterfly? How long did some ancestral pulp hang round inside a chrysalis (that came from who knows where) until a suite of chance mutations magically (how else?) re-programmed 'mush' into the concept of winged flight? How, in fact, did pupal death rise straightway (not even in a generation) to a form by which type-butterfly appears? One must ask also how, without the benefit of plan to reach anticipated adult form, serial immature phases took hundreds of millions of years to accrete. *It is noted that before any organism could 'create' any new stage of development it would have first to reach its present stage's adult limit then add that new stage.* This is because it has to reproduce to create the fresh, 'advanced' offspring. A reproducing caterpillar? Such back-to-front order is patently absurd.

From principle to practice, conceptual information to biology - imaginal discs, homeobox genes and other features simply demonstrate the *top-down* venture that, at climax, claps on stage a butterfly.

Butterflies are symbols of nemesis. At the clap of silent, fragile wings Darwinism logically dies; a giant is slain and at the same time flutter flags of life's innate, original intelligence.

Signs of obvious anticipation always flatten evolution. They squeeze the theory's time to death and thus compress it to impossibility. How can forethought be by chance? When is a plan not a plan? Is concept the same as lack of it? Biological evolution entirely *lacks* concept or target yet its theory equally lacks any serious explanation how, through many specified and integrated stages serial *and* parallel, the evolution of development '*must have*' occurred. Is this 'rationalism's' finest hour? **The fact is that all instances of metamorphosis (of which development in general is one) are another Darwinian black box.**

Sex, metabolism, metamorphosis and morphogenesis are four anticipatory and therefore conceptual processes. **The fact is that the evolution of development is as much a black box as the evolution of any of them. Their bio-logic hammers nails squarely into at least four corners of Darwinism's and thereby materialism's coffin.**

Chapter 6: Community

The system of Natural Dialectic is, as much as being an abstract reflection of the way things are, an application program. This program generates a template for both involuntary, instinctive and voluntary, chosen behaviours. It organises the pattern not only of 'hard science' and biology but politics, law and religion; it guides the aspirations of education; it is hard-wired into the humanities and, as such, is expressed in the very fabric of individual and social life. Its consequences, especially moral and psychological consequences, involve everyone. How?

We're going to work towards a Unified Theory of Community.[75] **From a bottom-up perspective humans are animals and mind evolved as a strange function of brains. Top-down,** *however, the first compass of community is universal and absolute.* It reflects a structure of creation whereby all things, psychological and physical, descend from an Absolute Source. In this sense alone is everything connected or, as some aver, 'is one'. **Such Absolute Community of Essence and Existence is illustrated in this diagram.**[76]

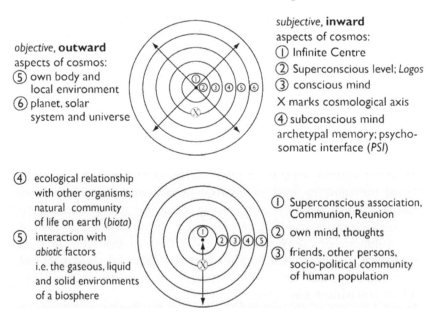

objective, **outward** aspects of cosmos:
⑤ own body and local environment
⑥ planet, solar system and universe

subjective, **inward** aspects of cosmos:
① Infinite Centre
② Superconscious level; *Logos*
③ conscious mind
X marks cosmological axis
④ subconscious mind archetypal memory; psycho-somatic interface (*PSI*)

④ ecological relationship with other organisms; natural community of life on earth (*biota*)
⑤ interaction with *abiotic* factors i.e. the gaseous, liquid and solid environments of a biosphere

① Superconscious association, Communion, Reunion
② own mind, thoughts
③ friends, other persons, socio-political community of human population

However, the Relative Community of Existence from a Standpoint of X (circles inward and outward from X) recalls the cosmological bearings by which we first oriented ourselves in Chapter 1.[77] From this point X, however, social circles radiate in *both* directions.

Internally there exist the symbolic yet most real relationships of mind. They involve other persons, bio-forms, events, objects and the paraphernalia of a life spent learning. This is the experience of your mind-world.

Externally you descend from X through physical interaction with your environment. The rings spread first through your nervous system (which is the closest physical associate of mind) to the rest of your skin and bones and their neighbourhood. This is your space-time location. It includes other biological bodies to which you physically and, more or less easily according to type, psychologically relate. If you restrict these to the human type then you involve the dynamic of family, friends, neighbours, interest groups, nations or the international scene. You take a social part, which may involve institutionalised politics, law and religion, in your community.

More broadly, you can think of life-forms (biotic factors) in terms of individuals, populations or communities of different kinds of organism. This positive, inclusive perspective is called ecological. *Thus ecology is about your 'wider body'.*[78] Such body includes a community of organisms (*biota*) that inhabit the surface region of our planet. This region, called an ecosphere, comprises living and non-living parts. The latter descend, as the rings convey, to include non-living elements (*abiota*) such as sources of energy, climate, ocean, water, mineral cycles and the soils of earth. The compound amounts to a stage on which, in different scenes, various players act. An ecological play is dynamic. In such a network the health and behaviour of each part affects its whole.

Firstly let's very briefly survey non-living factors.

We're here. Therefore at least our planetary zone is habitable. But life needs an endless supply of liquid water and thus strict, natural thermoregulation. Our thermal generator is a 'dwarf main sequence star'. Our lives are hung upon this lucky star. Precise strength and character of the four basic physical forces keep it burning radiantly. Not only earth's distance from the sun but also its unusually circular orbit slung on a fixed radius and a specific kind of rotation round its own axis is each exactly felicitous.

Pure energy (sunlight with harmful frequencies deflected or deleted) falls on gas. Earth's primary dynamic, sky, consists of a concentric suite of atmospheric shells. Each, like a membrane, offers its particular aegis to the life within. The living planet floats, egg-like in a white of air, within the warm, deep womb of solar influence. You might construe the stratosphere and ionosphere as buffers, membranes, even subtle skin.

Nested lower down the suite of air-light shells you find fluids of fertility - mists, rain, rivers and the oceans. These, lower atmosphere and ground water, assume the blood-like role of convectors, radiators and conductors. The amount of water on the blue planet has remained stable. So when rivers pour megatons of salts into the oceans how do these avoid acceleration into dead seas? How, also, has pH stability been crucially retained?

The Mother's bones, nails and hair are soils and solid, crustal rocks that, washed by storm and stream, yield minerals. These minerals life's producers, plants, absorb. Volcanoes throw up irregular formations like mountain chains whose various habitats permit an abundance of ecological niches. If life's sac is sky, the earth's skin is a crust of islands (or tectonic plates) that float on seas of magma.

Animate is coupled with inanimate. A system is a network of ideas or objects linked by common purpose. *By this definition life on earth is called an ecosystem.* An ecosystem includes living and non-living factors combined into a single self-regulating system. For James Lovelock's Gaia theory earth is a 'super-organism' made of all lives tightly coupled with air, oceans and surface rocks. Such a 'super-organism' maintains dynamic equilibrium. It comprises a totality that *seeks*, in a cybernetic manner, optimal conditions for life. Its variables include temperature, pH, salinity, electrical potentials etc. **'Cybernetic homeostasis', like 'program', is a conceptual phenomenon. Its presence in any machine indicates an underlying purpose.**

For their biological part individuals, populations or communities of different kinds of organism are involved. Life's geo-physiological health, the poise on which all ecosystems and their multicellular inhabitants depend, is in great part the gift of bacteria. Whatever their mode of origin, microbes of the kinds that exist today always existed. Tough and reliable, they toil relentlessly. They 'plough' the earth and continuously 'farm' organic substrates on which other organisms thrive. Bacteria might even, as the foundation of life's ecological pyramid, be construed as its primary, substantial, most important form of life; yet, working at the interface with inorganic matter, most 'robotic' too.

Provision, consumption, recycled waste. *You need all three.* Input, process, output. *Homeostasis needs all three.* Ecology is irreducibly, biochemically homeostatic and cyclical. Together every community and every level of life in each community cycles around each co-factor. Indeed, each organism plays one or more of the roles. You need three-in-one to peg the balance happily.

From positivity let's turn, before considering deliberate negativity, to nature's shows of seeming negativity. An ill wind, evil cold, cruel sea and other natural challenges (not least, inescapably, the body's own calamities) may threaten life with suffering. They may spell fearful pain and death. Such 'evil', as we understand, does not involve intent or animosity. It is, as matter is, oblivious: not immoral but amoral. Thus 'ill wind' does not blow with ill intent. It blows according to the fashion of inanimate design. In short, nature's so called 'negativity' is really its insensible neutrality.

Voluntary negativity is quite another thing.[79] Evil deals, intentionally, in lies and pain. Its condemnation isolates, its burden weighs you down. Negativity inflicted on another is the state of hell.

all below	Transcendence
lesser sorts of being	Supreme Being
vectored deeds	Potential/ Poise
balancing acts/ reaction	Balance/ Pivot
range	Super-State
range of shadows	Essential Light
moral spectrum	Good
↓ negative act	positive act ↑
division/ demonization	unification
from Truth	towards Truth
from Peace	towards Peace
body/ self-centred	soul-centred
passion	detachment
contraction	expansion
malevolence	benevolence
hate/ abuse	love/ care
crooked/ perverted	straight/ open
criminal/ demonic	saintly
a curse	a blessing
immorality	morality
descent/ darker	ascent/ lighter
negative wish	positive wish
tendency to create chaos/ cause pain	tendency to order/ pain relief
decline/ fall	lift/ helping hand
depriving	sharing
malefactor/ enemy	benefactor
evil/ sin	goodness/ virtue
darkness	light

Resistance, opposition and exhaustion - the world exists through shadows set against its light. Beneath transcendence truth is broken by polarity into opposing vectors; plus and minus are unbreakable a pair. But as free agents and not unconscious robots humans *choose* which way to act - and sometimes voluntarily create another's pain. Holistically-speaking, don't blame divine but human choice for evil. Thus, if evil's sourced in men's own heads, moral struggle isn't cosmic but just local.

Who denies that in a fallen world the bestial side of life is struggle for survival? 'Survivorship' and 'reproductive fitness' win a day that's governed by genetic products called your limbic system and its master glands. To rape, cheat, pillage and exploit is thus implicit in your naturally selected genes.

What, therefore, might a well-evolved and cunning despot's genes inflict upon 'the enemies of scientific reason'? Deceive, purge, pulverise? Eradicate all threats to his genetic egotism's article of faith - any-cost survival? Ruthless, forceful and, the best of all, efficient tyrants soon create such 'excellence' as Dante's hell.

From pain and isolation let us turn towards a Unified Theory of Community[80] regarding its sociological part.

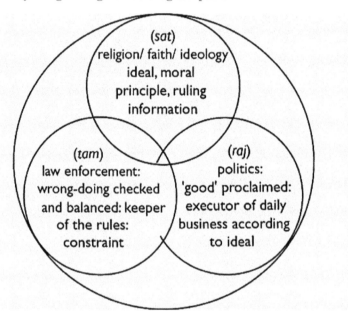

How best, philosophers enquire, live life? What solvent best dissolves the harshness of our problems on hard earth, what solution flows towards universal happiness? What regulates the dynamic equilibrium called peace? Does cosmic law (polarity) provide an answer even if it's not what *ego* might desire?

Down-to-earth solutions to the problems of disorder, ignorance and suffering involve two species of utopia - one objective, outward social and the other inward, personal and subjective. *We'll deal with the outward, large-scale solution first and then, as was the plan, work inwards towards a microcosmic, personal Nuclear Solution.*

Firstly, therefore, in considering a **Unified Theory of Religion, Politics and Law,** we speak about '*external association*'. Such association comprises relationships between an individual and his circumstance. It includes both positive and negative interaction with non-living objects, any living organism and, closest, other humans. At the heart of human community are religious, political, legal, educational, economic and personal communications with each other. The strongest, most important relationships are often with those in close physical proximity. This includes one's own body (habitually thought of as oneself), family, neighbours, colleagues and workmates. Of this group the closest are those 'on the same wavelength', that is, perhaps family but certainly friends. *What, therefore, should the instruments of social cohesion - law, politics and religion - most attentively and sensibly nurture but family and, within the family, its individuals. Is this your society's case?*

Religion, politics and law - this homeostatic triplex guides the individual and, by extension, his society. Is not the foremost regulator of the three, from which ideals derive, world-view? Secular or, if immaterial precedes and governs the material state of man, non-secular religion sets ideals. Ideals, as far as realised, compose desired utopia. In this case from Natural Dialectic's point of view best principle derives from Top Ideal. *Society's best formal answer is one that addresses evil issues from the State of Immateriality; our material solution is resolved, most definitely without recourse to science and without a test-tube anywhere in sight, by institutions charged with primary exercises in morality. Morality steps centre stage.*[81] Its values aren't of scientific kind but it is for sure the bigger player. The purpose of the agencies of social order - religion, education, law and politics - is to combat imperfections, minimise all criminal behaviour and promote relationships. *The whole point of morality is maintenance of individual and thence social balance and, wherever necessary, restoration of dynamic peacefulness.*

	body/ mind	*Soul*
	lesser selves	*Self*
	personae	*Psyche*
	concentric rings	*Nucleus*
	relativity	*Truth*
	relatively sane	*Sane*
↓	*external*	*internal* ↑
	descent	*ascent*
	exit	*return*
	lower self	*higher self*
	from Creator	*to Creator*
	from Centre	*towards Centre*
	from Goodness	*towards Goodness*
	criminal/ evil	*respectful/ good*
	insane	*sane*

Religion means 'a system of belief' and, whether vague or clearly defined, is therefore unavoidable. Whether atheistic or theistic, oriental, occidental or plain secularly 'liberal' it frames relationship with cosmos.

For materialism there is only one way. There is no direction but its own wherein all life is evanescent and thought dances, like will o' the wisp, inexplicably upon the waters of oblivion. Matter has *per se* no purpose. All things, including brain, are made of matter; this non-conscious factor has no will; no will, no choice, no morality. If this is hopeless and a counsel of despair then grow up, the creed exclaims, grow hard and strong because that's just the way this harsh, unfeeling, aimless and amoral cosmos is.

If, on the other hand, the mind-meat equation is delusional (Lecture 3), then mind is immaterial. In a 3-step cosmos it is *not* non-conscious. Purpose, attitude and relative free will make choices. 'Good', 'bad', 'right' and 'wrong' exist; law and morality are real.

And yet from what Reality do they derive? Is there, Centre Stage, an Absolute Morality? If, over mind, you treat with Nuclear First Cause, you treat with Metaphysical Ideal, Transcendence and Top Teleology; and with Free Will whose nature, Love, is paradoxically and by degrees constrained by various forms of mind. These forms are relative impurities.

	existence	*Essence*
	spectrum	*Source*
	hierarchy of riders	*Ideal*
	lower principles	*First Principle*
	consequent development	*Nucleus*
	belief	*Knowledge*
	religious faith	*Super-religious Fact*
	world-tree	*Archetypal Seed*
↓	*darkness*	*light* ↑
	peripheral formalities	*approach to nuclear core*
	bondage	*freedom*
	lack of intelligence	*intelligence*
	rigid adherence/ dogma	*thoughtfulness/ debate*
	hypocrisy	*sincerity*
	rule by fear	*rule by wisdom*
	arrogance/ intolerance	*toleration*
	wrong	*right*

Mind's eye looks down upon a world of soul-less things but, swivelled upward towards the Apex, seeks illumination's jewel.

You step up; and, in communion with Essential Soul, are transformed into supreme and living paradox. This we call nuclear religion. [82]

However, mankind's peripheral religions, the world's formal faiths, are social encrustations born of ritual surrounding nuclear experience. Saintly teaching is expressed in differing, local ways. Holy scriptures, creeds, imaginations, theologians, uniforms and prayer-houses represent an illumination-inexperienced congregation. At best this expression points upward towards a knowledge of the Peak Ideal. At worst, however, creeds breed isolation. An oft-bellicose delusion of any particular organised religion is that it's 'the only way' or 'sole repository of truth'.

For a peripheral religious believer explanation of his faith is not required; and for what dogma forbids none will suffice. For atheistic 'infidels' the religious delusion is that metaphysic has existence. For theistic believers various delusions stem from encrustation of ideals, misunderstandings, misinterpretations, fanatical conformities and their hypocrisies. 'My gang, your gang' mentality directly leads to conflict and, destructively, to war. Illumination is reversed. Ignorant perversion of the upturned approach to nuclear core is twisted round and downward into politics of fear and clerical control - the very opposite of Truth.

From ideal theory turn to the lower truths of egotistical pragmatics. Below the third-eye's balcony observe life's market-place. Survey the action, business and economy of body - seething crowds, cross-currents of humanity; turbulence of social ocean, individuals jostling with each other in a way that keeps (or fails to keep) the peace. What is a crowd but a community, communities a town and towns writ large across a wedge of continent? Individuals in society, men in a collective state, states dealing with each other in a way that keeps (or fails to keep) the peace - this is the cauldron, human politics, into which birth throws us and we boil. As with biological so with legal and political dynamics. **Resilient balance minimizes stress so that equilibration is, for body politic, the basis of its politics**. The purpose of a government is steering towards its stated goals. Are these not, in a human nutshell, education and, in providing for a healthy body, keeping healthy peace? *Its purpose is to protect its subjects' sense of equilibrium by maintenance of law and order in its realm.* Employing metaphysical ideal how best, philosophers enquire, to legislate? Aligned with which criteria might a leader govern beneficially?

Are conscience, self-control and an internal law to be preferred? Or don't they count and you invite a visit from the officers of law? Rules leash animals with reason; they frame your practices with principle. You are constrained within a mental box of regulation. If you flout the ideals that it represents then an external leash of conscience must restrain. Those officers detain and teach you with the pain of punishment. They might even lock you in a hard box called a cell.

The second of two species of utopia is inward, personal and subjective.[83] *We now work inwards from Point X towards the Real Deal, Nuclear Seed, Nature's Inward Positivity.* No crime! No prisons, courts or punishments! Do such societies, outside monastic, still exist in which self-government eliminates the reckless, disrespectful element? Where, really, does self-government exist but mind? So crimelessness has great potential. Nature's Inward Positivity exists in every human but its exercise is voluntary and needs choice.

Delve further in. Take nuclear religion to your own (and everyone's) extreme. *Internal, individual association involves just one relationship.* How well do you live within yourself? How to realise The Real Deal - Your Self? As far as possible return to Origin.

Nuclear Religion is about (↑) return from the periphery of creation to its Natural Centre. Isn't the Aspiration of a human life, naturally inlaid but much neglected and distracted, to distil the mind's pollution to a pure distillate and thus, as every mystic always told you, reunite your life with Life? Essential Psychological Unification: Return to Source.

existence	*Essence*
relative truth	*Truth*
duality	*Unity*
↓ *division*	*unification* ↑
material focus	*immaterial focus*
towards illusion	*towards Truth*
no thanks/ ingratitude	*thanks/ gratefulness*

In conclusion, I ask whether *information* may not be the hidden, immaterial factor that causes currently materialistic paradigms to shift. Whatever the case, I hope that you have enjoyed the exploration of a philosophical architecture whose *motif* is the binary pattern of Natural Dialectic. We have, considering both inanimate and animate, circled round the universe. ***What is, from the evidence, the final line and highest of conclusions? What, at the most natural core of cosmos, is the nature of Truth? The choice, the faith is at the end between material chance and an Informative Creator.***

Glossary

adaptive potential: involves possible changes to super-coded switches and recombinant refinements intrinsic in the genomic program of any particular biological type (*SAS* Chapter 23).

anti-entropy: *see* **negentropy.**

archetype: basic plan, informative element; conceptual template; pattern in principle; instrument of fundamental 'note' or primordial shape; causative information in nature; 'law of form'; nature's script; Natural Dialectic's 'holographic' edge, omnipresent but invisible because it's metaphysical; the place, in lower case, where metaphysic and its physic meet; morphological attractor or field of influence in universal mind; prototype-in-mind (maybe related to Platonic ideas, Aristotelian entelechies and/ or Jungian archetypes); potential matter seen as hard a metaphysical reality as, say, particles are physical realities; program(s) naturally stored in cosmic memory - simple in terms of inanimate physical 'law' (of particles and forces), complex in terms of animate structure/ function/ behaviour; information stored in a typical mnemone; in biology, metaphysical correlate of biological type/ super-species that is physically expressed in code as *potential* form; abstract or metaphysical precursor; as thought is father to the deed or plan is prior to ordered action, so archetypes precede physical phenomena; pre-physical initial condition of matter; see also *SAS* Chapters 16, 17 and 19; also *PGND* Chapters 11 and 13-15.

ATP: Adenosine TriPhosphate, life's standard bearer of chemical/ heat energy; a cell's agent of energy transmission; a biological 'match' or 'battery'; an active cell may discharge many thousand units of *ATP* per second to drive its metabolic machinery; these are recharged by respiration; *ATP* also plays a critical informative role in the transmission of nervous and possibly other signals.

big bang: *see* **transcendent projection**

black box: process or system whose workings are unknown.

chaos: a confusing notion with three main but disparate implications - emptiness, disorder and randomness; Greek word meaning chasm, emptiness or space; structureless 'profundity' that pre-existed cosmos; *prima materia*, or primordial energy structured by regulation of divinity archetype or natural law; anti-principle of cosmos i.e. disorder; any case of actually or apparently random distribution or unpredictable behaviour; also apparently random but deterministic behaviour of systems (e.g. weather, electrical circuits or fluid dynamics) sensitive to initial conditions.

chemical evolution: also called abiogenesis, biopoesis, chemogenesis or prebiosis; implies that lifeless chemicals 'evolved' to the point whence they could 'self-construct' the primary unit of life, a reproductive cell; it means the generation, perhaps gradually over a long period of time, of life from non-living components by physical means alone. This process is integrally part of, strictly not the same as, Darwin's consequent evolution.

chloroplast: organelle in plant cells containing photosynthetic apparatus.

chromosome: a 'book' in the 'encyclopaedia' of life; the human genome contains 46 chromosomes.

code: the systematic arrangement of symbols to communicate a meaning; code always involves agreed elements of morphology (the form its symbols take), syntax (rules of arrangement) and semantics (meaning/ significance); without exception such prior agreement between sender (creator/ transmitter) and recipient involves intelligence.

Communion: the Christian term for mystic union (see also transcendence, *samadhi*, *nirvana*, holy grail); realisation of Universal Truth; Salvation; Absolution; core subjective experience; attainment of Supreme Being; Core Psychological Experience.

conscio-material dipole: illustrates the basic components of polar existence; informative and energetic components may be graphically modelled (as *fig.* 2.5) on vertical y- and horizontal x-axes respectively; the origin of such graph (zero information and zero energy) represents the cosmic sink - an abyss of space); sources of the couple extend from infinity towards this sink; Archetypal Potential scales from Psycho-Logical, Informative First Cause through grades of mind to non-conscious zero (matter); and the projection of potential matter (first cause physical) drops from original concentration of ultra-heat to, again, zero; thus cosmos is viewed as the gradual embodiment of Uncreated Source; it represents a scale of possibilities expressed as typical yet individual forms; by this token embodied soul is subject to individual incorporations (psychological and physical); these relatively dynamic forms constitute its psychological and, in the biological case, bodily circumstance; Natural Dialectic simply models such a hierarchical description of polar creation by the use of spectrum, concentric rings and, step-wise, ziggurats (see also cosmos and creation below).

cosmic fundamentals: cosmic psychological and physical qualities; three basic states or tendencies; universal ingredients whose mixture is variously expressed in every object and event.

cosmological axis: human pivot; the point at which subjective and objective perception meet; eye-centre; third eye; thought centre; *ajna chakra.*

cosmos: physical universe, universal body; denotes orderly as opposed to chaotic process; involuntary pattern of nature; also equated, including metaphysical mind, with existence as a whole; seen, dialectically, as a projection through the template of metaphysical archetypes

creation:	origination; physical or psychological arrangement; mind creates with purpose, matter without; creation means active production but also passive result; a creation will have been informed by force of mind and/or matter.

D

dialectical stack:	stack of opposites; columnar expression of polarity; there are two kinds of stack - primary or non-vectored and secondary, vectored; primary (essential) stacks set (*Sat*) Unity against ($\downarrow$ *tam*/ *raj* $\uparrow$) duality (for elaboration see especially Chapter 2 and *figs*. 1.4, 2.2, 24.1 and 24.2); secondary (existential) stacks represent the various kinds of polarity from which the changeful web of existence is composed; each pair of polar 'anchor-points' implies a scale or dynamic range that runs between 'paired opposition' or 'complementary covalency'; stacks do not necessarily list synonyms or make equations; ***their perusal is intended to promote connections because consideration of connections tends to help unify/ collate/ organise one's working comprehension of any matter in hand.***
DNA:	a complex chemical; a large bio-molecule made of smaller units, nucleotides, strung together in a row; a polymer in the form of a double-stranded helix; a medium superbly suited to the storage and replication of 'the book of life'; 'paper and ink' on which the genetic code is inscribed; an organism's 'hard drive'; *DNA* is such an elegant, efficient and densely-packing form of information storage and manipulation that *DNA* computing by humans is now a fledgling technology; in this fast developing field silicon- based technologies are replaced using *DNA* and other bio-molecular hardware.

E

electromagnetism:	physics of the field that exerts an electromagnetic force on all charged particles and is in turn affected by such particles: light/ e-m radiation is an oscillatory disturbance (or wave) propagated through this field; light; light paradoxically involves a perfect, polar balance between contractive/ magnetic and radiant/ electric components.
entropy:	a measure of the amount of energy unavailable for work or degree of configurative disorder in a physical system (see second law of thermodynamics); inertial aspect of an energetic, material or conscious gradient; diffusion or concentration gradient outward from source to sink; drop towards 'most probable' outcome i.e. inertial slack; a measure of disintegration or randomness; expression of the (*tam* $\downarrow$) downward cosmic fundamental; a major property of matter, closely coupled with materialisation; in a closed system, which the universe may or may not be, this tends the eventual loss of all available energy, maximum disorder and the exhaustion of so-called 'heat death'.
enzyme:	protein catalyst without whose type metabolism (and therefore biological life) could not happen.

epigeny: genetic super-coding; contextual punctuation; chemical modification of *DNA*; also extra-nuclear factors that may cross-reference with genetic expression.

equilibrium: three modes of equilibrium are (*sat*) balance of poise or pre-active potential; (*raj*) dynamic balance occurring in all regular cycles, wave-forms and cybernetic homeostasis that is basic to the stability of life-forms; and (*tam*) inertial equilibrium that results from diffusion of information or energy; it equates with exhausted inaction or 'flat', impotent rest; such post-active inertia represents the most probable distribution of energy/ matter with the least energy available for work viz. the most random arrangement permitted by the constraints of a system; expressed in psychological terms as ignorance, unconsciousness or sleep; see also equilibration, *karma* and *fig*. 3.3 'Pivoted Existence'.

Essence: (*Sat*) Supreme or Infinite Being; Substance (perhaps Spinoza's Substance) 'prior to' or 'above' existence; Pure Consciousness/ Life; Peace that transcends all psychological and physical action; the root of an essentially undivided universe; Uncreated One within which and whence all differences have their being; Apex of Mount Universe; goal of saints/ 'philosopher kings'; the 'point' at which All-Is-One.

eukaryote non-prokaryote; any organism except archaea, bacteria and blue-green algae.

evolution: there are today *four* main usages of this word; each 'loading' derives from the original Latin, 'evolvere', meaning to unroll, disentangle or disclose; the *first two*, physical and biological, are conceived as natural/ mindless processes; the *second, mindful pair* is of psychological/ teleological import; specious ambiguity may conflate or switch between the fundamentally separate pairs of meaning. *Firstly,* in the scientific context of physics and chemistry, the word is used to describe change occurring to physical systems; the laws of nature can't, it seems, evolve through time but stars, fires, rocks or gases can. *Secondly*, though also subject to the 'rules' of entropy, biological evolution is a theory of *random progression* from simple to complex form; it thereby implies increasing, codified complexity; while retaining the 'hard loading' of physical science it also, ambiguously, claims that codes, programs, mechanisms and coherent, purposive systems - normally the province of mental concept - self-organise by, essentially, chance; such confusion, the basis of naturalism, is compounded by failure to distinguish between, on the one hand, ubiquitously observed variation (called micro-evolution) and, on the other, Darwinian 'transformation' between different sets of body plan, physiological routines and associated types of organism - such 'black-box macro-evolution' as is never indisputably observed; to evoke a naturalistic ambience it is fashionable to use 'evolved' interchangeably with or to replace the words 'was created', 'was planned' or 'designed'; finally, it is noted that the coded, choreographed development of a zygote, packed with

106

anticipatory information, through precise algorithms to adult form is the absolute antithesis of blind Darwinian evolution. ***Thirdly***, man certainly evolves ideas; intellect can evolve 'purposive complexity'; we invent all kinds of codes, schemes and machines; we devise increasingly complex theories and technologies; and we evolve an understanding of natural principles; this, which all parties accept, is an informative, psychological sense of 'evolution'. The ***fourth*** sense of evolution, at least as near to the original Latin as the other three, is the spiritual usage; immaterial spiritual evolution, unacceptable to materialists and unknown to physical science, is at the very heart of holism; in this voluntary sense of evolution practitioners cast off material attachment, evolve and merge into the *Logos*; evolution (or, perhaps better, centripetal involution) of the soul is their great business; their aspiration is to unite with The Heart of Nature.

evolution pre-Darwinian: minority/ anti-mainstream pre-Socratic snippets and sense-based Epicureanism lionized by interpretations of post-18th century materialists; virtually undetectable eccentricity in Chinese, Indian and Islamic literature; natural selection treated by creationists al-Jahiz and Edward Blyth; Buffon, a non-evolutionist, addressed 'evolutionary problems'; Lamarck (evolution by inheritance of acquired characteristics); hints in poem by Erasmus Darwin.

evolution Darwinian: mechanism - natural selection; major tenets - common descent (inheritance), homology and 'tree of life'.

evolution: neo-Darwinian/ synthetic: as Darwinian, except synthetic theory adds random mutation as the mechanism for innovation; also adds a mathematical treatment of population genetics and various elements (e.g. geno-centric perspective) derived from molecular biology.

evolution: post-synthetic phase: natural selection and random mutation are acknowledged as mechanisms insufficient to source bio-information; post-Darwinian evolution invokes mechanisms from hypotheses such as *NGE* (natural genetic engineering) and 'evo-devo'; holistic possibilities also address the origin of complex, specified and functional bio-information.

existence: which 'stands out' from background 'nothingness'; the apparently divided universe; seemingly disparate, finite things; all motion/ change/ relativity; all psychological and physical forms and events.

F

first cause(s): first cause is first motion in a previously undisturbed, pre-conditional field; normally considered the first impetus to a chain of events; such 'horizontal causation' is complemented by the 'vertical causation' of Natural Dialectic's hierarchical view of cosmos as explained in Chapter 2.

First Cause Psychological is Archetype, Potential Informant or (see Chapter 5: Top Teleology) *Logos*; attributes of this Primary Source and Sustenance of Creation include omnipresence, omnipotence and omniscience.

first cause physical is called potential matter or archetypal memory; as the secondary source of creation it precedes physical phenomena; as such it is, transcending physical appearances, metaphysical; this 'physical nothingness' is therefore, paradoxically, the source of everything composing astronomical cosmos; it consists of their internal being as opposed to external manifestation or their essence as opposed to quantum or bulk state appearances; its void, with respect to the presence of finite phenomena, appears infinite; 'holographic' attributes of immanent archetype, the primary informant of our non-conscious, energetic universe, include omnipresence and omnipotence; also check the Glossary for archetype and transcendent projection.

free will: free will, reflecting cosmos, occurs in three stages; first, in non-conscious, automatic matter, that is, the physical universe, there is none; second, in conscious mind freedom of will is relative; the degree of this relativity is a function of the type of an impression, that is, of the higher or lower quality of a memory, purpose, form of thought or feeling on mind's spectrum; third, the Essential Nature of Unconstrained Free Will is, paradoxically, 'constrained' by Transcendent Love.

G

gamete: sex cell with half of full genetic complement i.e. a single set of chromosomes.

gene: generally means a basic unit of material inheritance; section of chromosome coding for a protein; digital file; a reading frame that includes exons and introns; the old one gene-one protein hypothesis is incorrect; in fact, by gene splicing, a particular piece of *DNA* may be used to create multiple proteins.

genome: total genetic information found in a cell: think of the genome as an instruction manual for the construction and physical operation of a given organism.

genotype: the genetic constitution of an organism, often referring to a specific pair of alleles; the prior information, potential, plan or cause of an effect called phenotype.

gravity: in physics an attractive mass-to-mass force or warping of space-time; in Natural Dialectic the term is redefined more broadly - the agency of its (*tam* ↓) downward vector includes all psychological and physical factors of materialisation; such 'gravitational' factors and their properties are listed in the left-hand column of Secondary, Existential Dialectic; they include pain, pressure, confinement, strong nuclear force, mass, electromagnetic binding, inertia, entropy, 'standard' gravity and so on; gravity might be summarised as 'negative power' or 'the principle of death'.

H

holism: embraces material as well as immaterial (psychological) science in its compass; opposite of reductionism; the view that a whole is greater than the sum of its parts; the extra metaphysical (immaterial) ingredient is identified by Natural Dialectic as information; information implies the purposeful design, development and arrangement of contingent parts in a working system; cosmos may operate according to a Logical Norm.

hologram: a 3-d photograph made with the help of lasers; unlike a normal photographic image each part of it contains the image held by the whole.

homeostasis: vibratory or periodic control of a system to obtain balance round a pre-set norm; the mechanism of its information loop involves sensor, processor and executor; the operative cycle works by negative feedback; dynamic bio-equilibrium; psychological (nervous) and biological cybernetics; the informed basis of biological stability.

homeotic gene: gene involved in developmental sequence and pattern; high-level co-determinant of the formation of body parts.

I

illusion: is the cut between illusion and delusion an illusion? illusions, apparently outside the mind, appear real; a delusion, in it, we think real; illusion is a lesser truth; set against Absolute Truth (or Reality or Knowledge) it is a relative truth; hierarchical existence is composed of relative truths that range from slightly to completely false; only in Truth, only from the perspective of Knowledge do the illusions and delusions of existence wholly disappear (see Glossary truth; also *SAS* Chapter 4 and *PGND* Chapter 16: Truth, Appearance and Reality).

information: the immaterial, subjective element; information is action's precedent; informative potential is both a psychological and (by way of archetype) physical entrainer; information is the inhabitant of its own centre, mind, whose substrate is consciousness; *active* information knows, feels, purposes and codifies; it recognises meaning; on the other hand *passive* information reflects active; it is stored as subconscious memory; or is fixed in the expressions of non-conscious matter according, universally, to the archetypal behaviours of natural bodies or, locally, to particular constructions by life-forms.

informative entropy: loss of information due to degradation of its carrying medium; such a medium may be metaphysical (mind) or passive and physical (for example, computer files or genetic code); and its entropy may be metaphysical (loss of memory, focus or consciousness) or physical (for example, genetic mutation); the informative correlate of such degeneration is diminished organisational capacity, meaning or thrust of original purpose.

informative negentropy: gain of informative clarity; increasingly focused, purposive specificity; associated with knowledge, wisdom, grasp of principle and pristine construction.

inversion: turning upside-down or inside-out; reversing an order, position or relationship; in a hierarchical sense inversion is allied with the reflective asymmetry of opposite poles; information outwardly expressed; pole-to-pole reversal integral to dialectical structure; various kinds of inversion (cosmic and micro-cosmic (or biological)) are discussed.

K

karma: the law of balance between action and reaction; equilibration such as underlies all mathematical equation; a deed with implications of the reactions or 'payback' it provokes; fruit or result of previous thoughts, words and deeds; applies as rigidly to metaphysical (psychological) as, in Newton's Third Law of Motion and mathematical *equations*, physical events.

L

levity: agency of the (*raj* ↑) upward vector; dialectical converse of gravity; psychological and physical 'levitatory' forces lift or stimulate; they are listed in the right-hand column of Secondary, Existential Dialectic and include light, heat, excitement, dematerialisation, release, negentropy, focus of interest, affection and so on; physically, levity includes anti-gravity or the intrinsic property of matter's absence, space; generally summarised as 'positive power' or 'a buoyant principle of liveliness'.

logic: analysis of a chain of reasoning; principles used in circuitry design and computer programming; 'normative reason' relates to the basic axiom(s) of a given standard e.g. *bottom-up* materialism or *top-down* holism; three main logical thrusts are: (1) inductive (premises/ observations supply evidence for a probable/ plausible conclusion) as in the case of experimental science working *bottom-up* from specific instances to general principle: (2) abductive (best inference concerning an historical event): and (3) deductive (conclusion in specific cases reached *top-down* from general principle): two pillars of logic are holism and materialism; holism employs mainly deductive/ abductive operations and a Logical Norm; materialism tends to inductive/ abductive operations whose axis is non-conscious force and chance.

Logos: First Cause; Prime Mover; Causal Motion that sustains creation's conscio-material gradient.

M

macrocosm: the physical universe of astronomy and cosmology; dialectically, the whole of existence (i.e. both universal mind and universal body) as opposed to individual, microcosmic objects and events - including the human body.

macro-evolution: large-scale, non-trivial evolution; process of common or phylogenetic descent alleged to occur between biological orders, classes, phyla and domains; includes the origin of body plans, coordinated systems, organs, tissues and cell types; unexplained by mutation, saltation, orthogenesis or any known biological mechanism; sometimes called 'general theory of evolution' (*GTE*); macro-evolution, an extrapolation from Darwinian micro-evolution vital to sustain the materialistic mind-set, is conjecture.

matter-in-practice: bulk, bonded matter including all molecular-based substances; gross matter; external appearance.

matter-in-principle: quantum phase; particles and forces; subtle matter; internal cosmic drivers.

meiosis: shuffling the information pack: variation-on-theme; mechanism for the production of haploid gametes; genetic postal system for sexual reproduction.

metabolism: body chemistry.

metaphysic: = non-physical/ immaterial/ psychological/ unnaturalistic (if physic is equated with natural); physically expressed as specific/ intended arrangement/ behaviour of materials; physical behaviour reflects metaphysical blueprint; involves element of information; also involves symbol/ code/ abstraction/ logic/ reason/ mathematics; also message/ meaning/ goal/ teleology; also consciousness/ mind/ life/ experience/ feeling; and also morality/ force psychological/ emotion; involves innovation/ creativity/ art/ invention/ aesthetics.

microcosm: an entity that reflects the universe by containing all its basic constituents. Used especially of the human state where it may refer to both mind and body or, in a purely physical context, body alone.

micro-evolution: misnomer; non-progressive, small-scale variation within a species or, more broadly, between strains, races, species and genera; *variation/ adaptation within type*; trivial Darwinian changes that may occur by natural selection/ ecological factors acting on genetic recombination or mutation; sometimes called 'special theory of evolution' (*STE*), micro-evolution/ variation is a fact.

mitochondrion: organelle in eukaryotic cells containing the apparatus for aerobic respiration.

mitosis: conservative copying and delivery of genomes in cell division; genetic reprinting; genetic postal system for asexual reproduction.

mnemone: a division of memory: the two divisions are personal mnemone (likened to a working cache or data store) and typical mnemone (likened to a *ROM* or an operating system); typical mnemone is a synonym for archetypal memory; Natural Dialectic's definition involves no 'cultural' connotation whatsoever and is thus wholly distinct from evolutionary psychology's use of the word; typical mnemone is an umbrella word for natural memory; term also applies to each specific type of organism's archetype.

morphogene: part of typical mnemone or archetypal memory relating to physical construction; morphological attractor; the component of subconscious mind associated with electrochemical function and thereby body; morphogene is the dominant, perhaps exclusive, aspect of mind in unconscious organisms such as plants or fungi.

morphogenesis: the development of biological structure; more generally, the production of physical form.

mutation: accidental change to genetic code.

mysticism: quite different from objective, it is the subjective science; not philosophy, religion or opinion but practice to achieve communion with natural, inner, immaterial truth; esoteric as opposed to exoteric, materialistic discipline; 'science of the soul'; as gyms and physical action are to athletes so meditative exercise and psychological stillness are to mystics; involves psychological techniques to achieve a clear, rational goal - purity of consciousness and thereby understanding of the fundamental nature of the informative principle, mind; since life is lived in mind a mystic seeks consummate knowledge of life's source and sanctum, that is, communion with its deathless heart; adepts were, are and will be 'Olympian' meditative concentrators.

N

natural law: the automatic, reflex and mathematically describable behaviour of a physical entity; likewise the repetitive nature of its interactions with other entities.

natural dialectic: dialectic is a method of discourse between two (or more) people holding different points of view about a subject, who wish to establish the truth of the matter guided by reasoned arguments; such dialogue has been central to European, Buddhist and, in the Taoist treatment of opposites, Chinese philosophy since antiquity. In Europe it was made popular by Plato and Aristotle; also William of Ockham, Thomas Aquinas and latterly, with differing shades of usage, Hegel, Kant, Marx and others. At medieval Oxbridge it was taught under the heading of logic along with rhetoric and grammar; professors examined this threefold 'trivium' in dialectical discourse with their student sat on a three-footed stool (or tripos); if successful he was awarded a degree. This book's Philosophy of Natural Dialectic uses its own robust, binary vehicle with which to make comparisons, connections and explanations; its polarities represent not only human but the cosmic spine; or, if you like, the system generates a muscular body of philosophy that accurately reflects the order of the cosmos. It generates an abstract, metaphysical machine, tight-knit, well riveted by bolt and counter-bolt, the simplest working model of the universe.

naturalistic methodology: also known as 'methodological naturalism', this strategy is, strictly, not concerned with claims of what exists or might exist, simply with experimental methods of discovering physically measurable behaviours; thus only materialistic answers to any question (e.g. how biological forms arose or the nature of mind) are deemed 'scientific' or 'scientifically respectable'.

negentropy: opposite of entropy; lowering of entropy; expression of the (*raj*) upward-pointing cosmic fundamental closely coupled with stimulus, dissolution and dematerialisation; a measure of input, cooperation or synthesis; motive/ fluidising aspect of an energetic, material or conscious gradient; gain of energy, configurative order, information or consciousness in a system; when used in terms of information negentropy involves gain in order or understanding of principle from which different actualities derive; a measure of the amount of concentrated/ conceptual information, specific, intentional complexity or conscious arrangement in a system; a natural and essential property of mind.

non-existence:	where creation = formful existence, non-existence is formless; the polar opposite of physical space and time is Transcendent Potential; such pre- or super-existential formlessness is non-existent; Absolute Non-Existence is Essential; however relative non-existences of two kinds also occur; the first kind is metaphysical/ subjective and therefore psychological; it involves the absence of a specific psychological form or event; unconscious oblivion is one such non-existence; the second kind involves the local absence of a possible physical event (an object is a 'slow event'); impossibilities are non-existences but imaginations of non-existence (including symbolic abstractions, hypothetical entities, physical absences, absolute emptiness and the number zero) exist; furthermore, the nothingness of space and time, the zero-point of calculus and zero's empty set together constitute the basis of physical science and mathematics.
nucleic acid:	*see DNA* and *RNA*
nucleotide:	basic, triplex unit of nucleic acid polymer; monomer composed of phosphate and sugar (the 'paper' part) and base (the 'ink letter'); letters' of the genetic alphabet are (G) guanine, (C) cytosine, (A) adenine and (T) thymine. In *RNA* thymine is replaced by (U) uracil.

O

object:	a slow, although energetic, event; apparent fixation.
Om:	universal sound, fundamental reverberation, basic creative agent; sometimes spelt *Aum*, a Sanskrit word whose Semitic transliterations are Am'n, Amin and Amen; see also First Cause, *Logos*, *Kalam*, *Shabda* etc.
order:	regular, regulated or systematic arrangement; organisation according to the direction of physical law; passive information by which things are arranged naturally (with predictable but non-purposive complexity) or purposely (with innovative or specified complexity); mind, generating specified complexity in the order of its technologies and codes, actively informs; the orders of mind are meaningful, the orders of matter lack intent.

P

PAM, PAND, PCM **and** *PCND*:	philosophical gambits; see Primary Axioms and Corollaries.
photosynthesis:	process by which inorganic carbon is introduced to the biological zone and energetic sunlight fixed as a crystalline molecule of storage, a sugar called glucose.
phylogeny:	evolutionary history; relationships based on common or evolutionary descent.
potential:	poise; latent possibility; potent non-action that precedes any particular action or creation; in science potential energy is defined as the energy particles in a system (or field) possess by virtue of position/ arrangement; gravitational, electrical, electro-chemical, thermo-dynamical and other kinds of potential are recognised; in dialectical terms mind precedes matter, information precedes the pattern of material behaviour;

information is energy's pre-requisite potential; in this case *informative potential* involves two conditions; firstly, a pre-existential/ essential state of psychological potential; secondly, a pre-material, metaphysical fact of potential matter, archetype or laws of nature; if potential's pre-active equilibrium is related to the voltage of a full battery then aspects of psychological 'voltage', whose currents drive intentional behaviour, are purpose, will and plan; *see* Chapter 2 diagrams re. potential and first causes.

potential matter: see archetype.

Primary Axiom of Materialism (*PAM*): all objects and events, including an origin of the universe and the nature of mind, are material alone; cosmos issued out of nothing; life's an inconsequent coincidence, a fluky flicker in a lifeless, dark eternity.

Primary Axiom of Natural Dialectic (*PAND*): there exists a natural, universal, immaterial element - information; immaterial informs material behaviour; a conscio-material dipole that issues from First Cause informs and substantiates both mental (metaphysical) and physical creations; there is eternal brilliance whose shadow-show is called creation.

Primary Corollary of Materialism (*PCM*): the neo-Darwinian theory of evolution, that is, life forms are the product, by common descent, of a random generator (mutation) acted on by a filter called natural selection; such evolution is an absolutely mindless, purposeless process; the *PCM* is a fundamental *mantra* of materialism.

Primary Corollary of Natural Dialectic (*PCND*): the origin of irreducible, biological complexity is not an accumulation of 'lucky' accidents constrained by natural law and death; forms of life are conceptual; they are, like any creation of mind, the product of purpose.

prokaryote: non-eukaryote; bacterial type with little or no compartmentalisation of cell functionaries.

promissory materialism: belief sustained by faith that scientific discoveries will in the future justify/ vindicate exclusive materialism and, as a consequence, atheism; may involve a call to progress towards the technological provision of its 'promised land'.

protein: factor made from a specific sequence of amino acids to perform a specific task; 'informative' protein includes some hormones; skin, hair, bone, muscle and other tissues are made of 'structural' protein; 'functional protein' called enzymes mediates all stages in cell metabolism, that is, it catalyses all biochemistry.

***PSI* (psychosomatic interface)**: psychosomatic border; the level of mind-matter interaction; bridge between metaphysical and physical dimensions; potential matter; 'gap of Leibniz'; 'fit' of mind to matter; point of linkage between subconscious mind and non-conscious matter; gearing between instinct/ archetype and the behaviour of material objects and energies; as in the case of physical law, psychosomatic influence is both general in potential and local/ specific in engagement.

114

psychological entropy: a measure of loss of concentration, focus of attention or consciousness; loss of 'mental energy' or aptitude; the drop from waking to sleep; loss of knowledge, information or sensitivity; the gradient from intelligence through stupidity to oblivion; an expression of the (*tam*) downward cosmic fundamental in mind; a tendency predominant in lower, egotistical or selfish mind; increasing level of ignorance, anguish or immorality; loss of integrity, psychological disharmony or disintegration; see also *information entropy*.

psychological negentropy: a measure of gain in order; an increase in concentration, focus of attention or consciousness; gain in sense of purpose, 'mental energy' or aptitude; the rise from sleep to waking, 'dark to light' or unhappiness to happiness; gain in knowledge, information or sensitivity; the gradient of learning and spiritual evolution; an expression of the (*raj*) upward cosmic fundamental in mind; a tendency predominant in higher mind; increasing level of contentment, understanding and the natural morality of happiness; the ascent towards psychological radiance, harmony and integration. The converse of psychological negentropy involves *entropy of information*.

psychosomasis: operation across the psychosomatic border; mind/ body interaction; the one-way, morphogenic imposition of archetypal pattern on *physicalia*; the two-way exchange of information in sentient organisms through the agency/ medium of subconscious patterns.

Q

quantum: minimum discrete amount of some physical property such as energy, space or time that a system can possess; quantum theory states that energy exists in tiny, discontinuous packets each of which is called a quantum; an elementary discontinuity; an elementary particle e.g. photon or electron.

quantum level: matter-in-principle; 'internal', 'causal' or 'subtle' matter; the vibrant or energetic phase of physical organisation; zone of sub-atomic particles and forces; step (on cosmic ziggurat) between potential and bulk matter whose aspect is sometimes extended to include atomic and molecular interactions; small-scale substance underlying large-scale, sensible appearances.

R

raj: (↑) upward, levitatory or stimulatory cosmic vector.

reductionism: opposite of holism; the materialistic view that an article can always be analysed, split up or 'reduced' to more fundamental parts; these parts can then be added back to reconstruct the whole; a whole is no more than the sum of its parts.

religion: except in the case of materialistic species, a relatively temporary attempt to describe and interact with Essence; religions constitute changing metaphors for the indescribable, the transcendent; etymology of the word debated between Latin *relegere* (review) and *religare* (bind); this latter is the sense of *yoga* which also means joining, yoking or connection; *religio* means dutiful and meticulous observance; currently religion means world-view, mind-set or basic faith; whether of materialistic or holistic belief, it involves the non-negotiable substance of an individual or community's truth - notably as regards origins; antagonism between holistic practice and the naturalistic methodology of science is, because the couple deal with separate but complementary physical and metaphysical dimensions, flawed; a materialist/ atheist 'binds meticulously' to an evolutionary mind-set, a holist to either pantheism or a Primary Living Creator; in the case that self-deception is crucial to successfully deceiving others which, holism or materialism, is the religion that is ultimately true?

resonance: the tendency of a body or system to oscillate with a larger amplitude when subjected to disturbance by the same frequencies as its own natural ones; thus a resonator is a device that naturally oscillates at such (resonant) frequencies with greater amplitude than at others; resonance phenomena occur with all kinds of vibration, oscillation or wave; their sorts include mechanical, harmonic (acoustic), electrical (as with antennae), atomic and molecular.

S

sat: 'top' or essential cosmic fundamental; 'vector' of balance, neutrality.

science: Latin *scire* (know); knowledge; commonly understood as the practical and mathematical study of material phenomena whose purpose is to produce useful models of the physical world's reality.

scientism: a philosophical face of official, *de facto* commitment to materialism; today's majority consensus of what the creed of science is; an -ism born of *PAM*; a faith that all processes must be ultimately explicable in terms of physical processes alone; like communism, a one-party state of mind; a doctrine that physical science with its scientific method is ultimately the sole authority and arbiter of truth; a set of concepts designed to produce exclusively material explanations for every aspect of existence, that is, to colonise each academic discipline and build its intellectual empire everywhere; 'scientific fundamentalism' closely allied, when expressed in social and political terms, with 'secular fundamentalism', sociological interpretation of behaviour and the fostering of a humanistic curriculum.

secular fundamentalism: *PAM* as applied to the worlds of nature and of human society.

secularism: concern with worldly business; lack of involvement in religion or faith; secularism is generally identified, as defined by the dictionary, with materialism; for a secularist the ultimate arbiter of truth is human reason - ideas are open to negotiation so that even morality is relative; however many liberal agnostics, atheists and humanists argue that their metaphysical, philosophical system also embraces so-called 'universal' moral values and, as opposed to zealotry or the

	logic of evolutionary faith, a liberal politic of 'philosophical live-and-let-live'.
sub-state:	*opp.* super-state; impotence, discharge, exhaustion, final stage in the expression of potential; fixity; non-conscious base-state; state 'below/ subtendence; extreme negativity/ (*tam*) condition.
super-state:	potential; source of possibility; causal metaphysic/ archetype; state 'before' or 'above' subsequent expression; immanence; transcendence; precondition; (*sat*) priority.

T

tam:	(↓) downward, gravitational or inertialising cosmic vector.
teleology:	the doctrine that there is evidence of purpose in nature; doctrine of non-randomness in natural architecture; doctrine of reason ('for the sake of', 'in order to', 'so that' etc.) and intent behind biological and universal design.
third eye:	place where you think; point of metaphysical focus between and behind the eyebrows, that is, just above the physical eyes; HQ/ seat of mind beyond the sensory world; cosmological eye-centre; gate through which meditative concentration can pass; single way that leads within.

transcendent projection: creative projection from intrinsic Source; see Primary (Metaphysical) and Secondary (Physical) First Causes from Chapters 3 and 4; voluntary issue from potential mind or involuntary from potential matter; in other words, transcendent metaphysical projection is through Causal Archetype and transcendent physical projection through mnemone (see Index for references).

psychological: projection is through Causal Archetype called *Logos*, Holy Name, *Sat Nam*, *Om* and many other names. This projection is from Alpha Point, the first and highest cause of creation; see also *SAS* Chapter 5: Top Teleology and *figs. PGND* Chapter 13: First State of Super-consciousness.

physical: as Chapter 4 explains, physical first cause is projection through potential matter or base archetype in universal mind; such memory is the source of cosmo-logical language; it involves the orderly expression of energetic forces and particles; linkage, as Chapter 3 suggests, by psychosomatic resonance; emergence of physical phenomena from archetypal noumenon, that is, from unseen potential; an instantaneous 'miracle' that issues from 'within' non-conscious physicality; transcendently emergent, finely tuned expansion from 'inner' metaphysic into 'outer' material/ natural law; physical nothingness from whose from whose prior pointlessness (or 0-dimensional singularity) all points began; cosmic seed whence, *ex nihilo*, the world developed; projection whose appearance, once physical, is visible and perhaps described but certainly not explained by big bang theory; transcendent projection of archetype is possibly, to the constrained sensory and intellectual states of human mind, ultimately incomprehensible; its invisible dynamic, the practice of materialisation, may remain a fact beyond material understanding. for references involving more detail about psychological, physical and biological projections see *SAS* or *PGND* Glossaries.

biological: if matter is developed memory (*PGND* Chapter 14: Space) then Chapters 3 and 5, *SAS* Chapters 16 and 19 and *PGND* Chapters 13: Typical Mnemone and 15: Conceptual Biology expand this theme; biology is based on information in the form of code and codification, demanding forethought, is a product of mind; intrinsic archetypal program is reflected by extrinisic chemistry called *DNA*; in terms of Natural Dialectic *DNA* is passive information.

truth: what's correct or accurate; a universal truth substantiates, as source and sustenance, all things; man's holy grail is truth; *top-down*, within a hierarchically devolved construction (such as a machine, cosmos or other working system) truth is vested both in its source and physical appearance; in this case, viewed at their own level, psychological (subjective) or physical (objective) facts appear self-evident, obvious realities; however, they are relative or lesser truths when set against Truth Absolute; from different heights on a mountain slope the climber's perspective changes; so with Mount Universe, whose Peak represents Whole Truth; in the light of Peak Reality all existence seems real but is composed of relative illusions; it is a shadow play dependent on The Independent Source; man is born to seek and find this Essential Source, this Alpha Point, his Origin; see also Glossary: illusion.

U

unification: details are unified by their working principles, themes or programs; better to perceive intrinsic principle is to simplify or unify an understanding; progressive unification of forces is the grail of physics: Clerk Maxwell unified electricity and magnetism; electroweak or *GSW* theory brought in the weak nuclear force; now the goal is to include the strong nuclear force (*GUT*), gravity in a super-force and show that, in essence, particles and forces are interchangeable (super-symmetry and *TOE*); Natural Dialectic, also working with the maxim 'All is One', includes what sums to a hierarchical *TOP* or Theory of Potential; potential (see *SAS*: Glossary and Index for archetype and potential) is the absolute from which variant orders of relativity derive; the subjective potential for mind is consciousness and the objective potential for matter is archetypal memory; such archetypal element unites psychology with the physics of natural science; it is the informative precondition of physical and biological form.

V

vector: existential dynamic; direction of travel with respect to model or secondary stack used in Natural Dialectic; fundamental vectors ($\uparrow$ and $\downarrow$) denote relative gain or loss of information or energy; and, similarly, motion towards and from the axis, peak or source of cosmic model; general, metaphorical rather than

118

specific use of the word; not, therefore, the same as that defined by physics or biology.

Z

ZPE: zero-point energy; quantum vacuum; vacuum energy of all fields in space; residual energy of all oscillators at $0°K$; concept first developed by Albert Einstein and Otto Stern; intrinsic energy of vacuum; the ground-state minimum that any quantum mechanical system, in particular the vacuum, can have; remainder, according to the uncertainty principle, when all particles and thermal radiation have been extracted from a volume of space; residual non-thermal radiation; irreducible 'background noise'; 'quantum foam'; the potent, microscopic side of quantum vacuum (as opposed to impotent, macroscopic vacuum left by the apparent lack of anything); subliminal 'rumblings' of immaterial weak, strong and electromagnetic fields (called *ZPF*s); seething, jostling ferment of subliminal waves and particles in emptiness; a flux of unobservable 'virtual' matter and anti-matter that may or may not appear as the basis of observable forces such as electromagnetism, charge and perhaps inertial mass and gravity; a subtle facet of levity; the anti-gravity of dark energy (or the cosmological constant) has been postulated as a component of *ZPE*; suggested 'mother-field' support for electron orbits, atomic structure and thus the phenomenal universe.

zygote: fertilised egg.

Connections

SAS   Science and the Soul
A&E   Adam and Evolution
PGND A Potted Grammar of Natural Dialectic
AMA? A Mutant Ape? The Origin of Man's Descent
SPFP   Science and Philosophy - A Fresh Perspective

---

[1]   *SAS*: Chapter 0: Anti-parallel Perspectives; *PGND* Chapter 1: Primary Assumptions; *A&E*: Chapter 2; *SPFP* Lecture 1: Materialism and Holism.

[2]   *SAS* Glossary; *PGND* Chapter 1: Primary Assumptions; *SPFP* Lecture 1: Primary and Secondary Axioms.

[3]   *SAS* Preface; *PGND*: Chapter 2.

[4]   *SAS*: Chap. 1 Natural Dialectic's ABC; *PGND*: Chap. 3 Models; *SPFP* Lecture 1.

[5]   *see* Glossary.

[6]   *SAS*: Chapter 1 Natural Dialectic's ABC/ Glossary/ Index; *PGND* Chapter 4; *SPFP* Lecture 1.

[7]   *SAS*: Chapter 1 Natural Dialectic's ABC/ Index: stack and dialectical operator; *PGND* Chapters 5 and 6, also Index: stack and dialectical factor; *SPFP* Lecture 1: basic grammar.

[8]   *PGND Chapter 8: Essence; Chapter 10: existence; SPFP* Lecture 2.

[9]   *SAS* Chapters 8: Principles of a Unified Theory of Matter; and 9: Nothing.

[10]   *SAS* Chapter 8: Infinity and Index; an infinity of existential units can be mathematically defined; The Unity of Infinite Essence cannot.

[11]   *SAS* Chapters 2, 4, 8: appearances, relativities and lesser, seeming elements; *PGND* Chapter 8: essential characteristics in existence; *SPFP* Lecture 1: limitations of essential characters.

[12]   *SAS fig.* 2.4; *PGND* Chap. 8: positive and negative zeroes; Chap. 10 inversion; *SPFP* Lecture 1: inversion/ reflective asymmetry.

[13]   *SAS passim*; *PGND* Chapters 13 to 15: triplex sciences; *SPFP* Lecture 1: simple links and Lectures 2-6: triplex, hierarchical science.

[14]   *SAS* Chapter 18; *SPFP* Lecture 1; paradigm shift.

[15]   *SAS*: Chapter 0: Opposite Directions of Mind and Two Pillars of Faith; *PGND* Chapter 1: Two Pillars, A Dialogue of Faith.

[16]   *SAS* Chapter 2: First Principles; *PGND* Chapter 10: Existence.

[17]   *SAS* Chapter 3; *PGND* Chap. 7: A Hierarchical Perspective; *A&E*: Chap. 3; *SPFP* Lecture 2.

[18]   *SAS* Chapters 3: Transcendence, 16 *passim*, 17 *passim* and 19: Conceptual Biology; *PGND* Chapter 13: H. archetypalis; *AMA?* Chapters 24 and 25; *A&E* Chapters 7 and 8; *SPFP* Lectures 2-5; also Glossaries and Indices in the books mentioned.

[19]   *SAS* Chapters 3 and 14: Subtendence and Transcendence; *PGND* Chapter 9; *SPFP* Lectures 2 and 3.

[20]   *SAS* Chapter 2: Causation and *fig.* 6.1; *PGND* Chap 10; also *figs.* 11.1 and 13.5 horizontal (energetic) and vertical (informative) vectors of causation; *SPFP* Lecture 2.

[21]   *SAS fig.* 11.1; *SPFP* Lectures 2 and 3.

[22]   *see* Glossary; also *SAS* Chapter 5 and *PGND* Chapters 11 and 12; *SPFP* Lecture 2.

[23]   *SAS* Chapter 5: (*Sat*) Potential Information and Chapter 6; *PGND* Chapter 12.

[24] *SAS* Chapter 6: Information's Infrastructure - Code; also Chapter 20: *DNA*; also Glossary and Index; *PGND* Chapters 12 and 15: Nuclear Super-computing; *A&E* Chapter 2; *SPFP* Lectures 2 & 5.

[25] *SAS* Chap.6; *PGND* Chap. 13 harmonic oscillation, resonance; *SPFP* Lects. 2 & 3.

[26] *SAS* Chapter 6: Machines, also Mind Machines (Computers*): SPFP Lecture 2.*

[27] *SAS* Chapters 5 and 14*; PGND* Chapters 12 and 13*; SPFP* Lectures 2 and 3.

[28] *SAS* Chapter 5: Top Teleology, Chapter 13: Psyche and Psychology, Glossary, Index; *PGND* Chapter 10: Causality and Chapter 13: First State; *SPFP* Lectures 2 and 3.

[29] *SAS* Chapter 0: Delusions, Chapter 3: Hierarchical, Triplex Construction of the Cosmic Pyramid and Chapter 13: Neurological Delusion; *AMA?* Chapters 2 and 3; *PGND* Chapter 13: Neurological Delusion; *SPFP* Lecture 3.

[30] *SAS* Chapter 13: Psyche and Psychology.

[31] *SAS* Chapter 13: The Neurological Delusion.

[32] *SAS* Chapter 13: Consciousness; *PGND* Chapter 13.

[33] *SAS* Chapter 14: States of Mind; *PGND* Chapter 13: Consciousness and *Top-Down* Psychology; *SPFP* Lecture 3: *Top-Down* Psychology.

[34] *SAS figs.* 13.2 and 13.3; *PGND* Chapter 13: Does Brain originate or Mediate? *SPFP* Lecture 3.

[35] *SAS* Chapter 13: Build Yourself a Brain.

[36] *SAS* Chapters 15 and 16; *PGND* Chapter 13: Third Sate of (Sub-) Consciousness.

[37] *SAS* Chapters 16, 17 and 19; *PGND* Chapters 13 and 15; Glossary and Index in these and also in *A&E* and *AMA*; *SPFP* Lecture 3: Third State.

[38] *SAS* Chaps. 15 and 16: Psychosomatic Linkage/ Psychosomasis; *PGND* Chapter 13: Psychosomatic Linkage; *SPFP* Lecture 3.

[39] *SAS fig.* 15.5.

[40] *SAS figs.* 15.3, 4, 5 and Chapter 16; *PGND*: Chapter 13; *SPFP* Lecture 3.

[41] *SAS* Chapter 16; *PGND* Chapter 13: *H. electromagneticus.*

[42] *SAS* Chapter 16: Synchromesh 2 - Psychosomasis; *PGND* Chapter 13 last three sections; see also Glossary: resonance and indices: resonance, harmonic oscillation and vibration; *SPFP* Lecture 3..

[43] *SAS* Chapter 17; *A&E* Chapter 17; *AMA?* Chapter 24; *PGND figs.* 13.7 and 15.1; *SPFP* Lecture 3.

[44] *SAS* Chapter 8: Holy Grails; *PGND* Chapter 13: First State; *SPFP* Lectures 3 and 4.

[45] *SAS* Chapter 10: Grades, Principles and Times.

[46] *SAS* Chapter 7: Lady Luck and Lord Deliberate *passim*; *PGND* and *SPFP* see Glossaries: chaos, Indices: chance.

[47] https://evolutionnews.org/2010/04/roger_penrose_on_cosmic_finetu/

[48] *SAS* Chapter 9.

[49] *SAS* Chapter 10.

[50] *SAS* Chapters 11 and 12; *SPFP* Lectures 2 and 3.

[51] *SAS* Chapters 11and 12; *PGND* Chapter 8; *A&E* Chapter 16; *SPFP* Lecture 4.

[52] https://en.wikiquote.org/wiki/Max_Planck

[53] *SAS* Chapters 19 and 20; *SAS* Glossary and Index: code; *A&E* Chapter 4: Genes and Genesis; *AMA?* Chapter 22; *PGND* Chapter 15; *SPFP* Lecture 5.

[54] *SAS* Chapter 23: Super-codes and Adaptive Potential

[55] *SAS* Chapter 20: Alchemy; *PGND* Chapter 15: Chemical Evolution?; *A&E* Chapter 9; *SPFP* Lecture 5.

[56] *SAS* Chapters 5, 6 and 19; *PGND* Chapter 15; *SPFP* Lecture 2.

[57] *SAS* Chapter 19: The Central Executive is Homeostasis; Index: balance and equilibrium; see also *PGND* Chapter 15; *SPFP* Lecture 5.

[58] *SAS* Chaps 5, 6, 14, 17, 19, 23, 25; *PGND* Chaps 11-13.

[59] *SAS* Chap. 21: Energy Metabolism Perchance?; *A&E* Chap. 10: Life's Engine; *SPFP* Lecture 5; since chemosynthetic systems are different yet as complex as photosynthetic ideas that convergence, coevolution or that the latter might have taken over from the former are, since they do not explain system origin, fanciful.

[60] *SAS* Chapter 19: Nuclear Super-computing and Conceptual Biology; *SPFP* Lecture 5; also check Indices: switch.

[61] *SAS* Chapter 23 and Glossaries.

[62] *SAS* Chapter 22: *A&E* Chapter 3: Hierarchy; *SPFP* Lectures 2-5.

[63] *SAS* Chapter 22: The Origin of Secies; *AMA?* Chapter 9: When is a Man a Man?; *SPFP* Lecture 5.

[64] *SAS* Chapters 20 and 21; *PGND* Chapter 15: Chemical Evolution?; *A&E* Chapters 4, 8, 9 and 10.

[65] *SAS* Chap. 22: Tree of Life; also 'genetic discordance' and 'anomalous gene trees'; *SPFP* Lecture 5.

[66] *SAS* Chap 22: Fossils; *A&E* Chapters 13 and 16; also Meyer: Darwin's Doubt..

[67] *SAS* Chap. 22: The Editor; *A&E* Chapter 5: Sports, Survival and the Hone; *SPFP* Lecture 5.

[68] *SAS* Chapter 23: The Creator; *A&E* Chapter 5; *PGND* Chapter 15: What's the Problem?; *SPFP* Lecture 5.

[69] *SAS* Chapters 21, 23 and 25; *A&E* Chapters 2 and 9; *PGND* Chapter 15; *SPFP* Lectures 2 and 5 esp. concerning development.

[70] In the 'Origin of Bio-information and the higher taxonomic categories' (Proceedings of the Biological Society of Washington 4-8-2004) S. Meyer argues that no mechanistic theory can account for the amount of information needed to build novel forms, systems or organs of life.

[71] *SAS* Chapters 24 and 25; *A&E* Chapters 7 and 8 for more detail as regards Reproduction and Reproductive Archetype; *SPFP* lecture 5..

[72] *SAS* Chapter 24; *A&E*: Chapters 7 and 12; *SPFP* Lecture 5.

[73] *SAS* Chapter 24; *A&E* Chapters 7 and 13.

[74] *SAS* Chapter 25 *passim*; *A&E* Chapter 8; *SPFP* Lecture 5.

[75] *SAS* Chapter 26; *SPFP* Lecture 6.

[76] Absolute Community of Essence and Existence: see also *SAS* Chapter 4 and *PGND* Chapter 16: Truth, Appearance and Reality.

[77] *SAS fig.* 3.5; *PGND fig.* 7.4; *SPFP* Lecture 2.

[78] Ecology: *SAS* Chapter 26.

[79] *SAS* Chapter 26; *SPFP* Lecture 6.

[80] Unified Theory of Community: Social and Individual Parts: *SAS* Chapter 26.

[81] *SAS* Chapters 4 and 26; *PGND* Chapter 26; *SPFP* Lecture 6.

[82] 'Religion' is derived from the Latin 'to bind back'; 'yoga', equally, means binding (to a discipline); Nuclear Religion; see *SAS* Chapter 5 and 26: Top Teleology; *PGND* Chapter 13 Transcendence; *SPFP* Lectures 2, 3 and 6.

[83] Individual Association: *SAS* Chapters 4, 26 and 27; *PGND* 16: Science to Conscience and Is There an Absolute Morality?; *SPFP* Lecture 6,

Index

130

T

U

V

The author has recently written a few more books (available from Amazon, Foyles, Waterstones, Barnes & Noble etc. and see website addresses on p.2):

Lightning Source UK Ltd.
Milton Keynes UK
UKHW020922070819
347548UK00004B/17/P